W9-BAV-451

BILLIONS AND BILLIONS

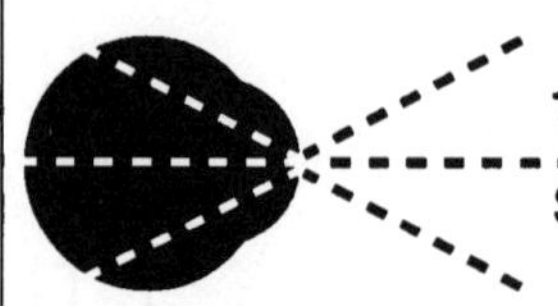

BILLIONS AND BILLIONS

Thoughts on Life and Death at the Brink of the Millennium

CARL SAGAN

Thorndike Press • Thorndike, Maine

Published in 1998 by arrangement with Random House, Inc.

Thorndike Large Print® Americana Series.

The tree indicium is a trademark of Thorndike Press.

The text of this Large Print edition is unabridged.
Other aspects of the book may vary from the original edition.

Set in 16 pt. Plantin by Minnie B. Raven.

Printed in the United States on permanent paper.

Library of Congress Catalog Card Number: 97-91345
ISBN: 0-7862-1363-9 (lg. print : hc : alk. paper)

TO MY SISTER, CARI,
ONE IN SIX BILLION

Contents

PART III

WHERE HEARTS AND MINDS COLLIDE

List of Illustrations

Part I

THE POWER AND BEAUTY OF QUANTIFICATION

Chapter 1

BILLIONS AND BILLIONS

There are some . . . who think that the number of [grains of] sand is infinite. . . . There are some who, without regarding it as infinite, yet think no number has been named which is great enough. . . . But I will try to show you [numbers that] exceed not only the number of the mass of sand equal to the Earth filled up . . . but also that of a mass equal in magnitude to the Universe.

ARCHIMEDES (ca. 287–212 B.C.),
The Sand-Reckoner

I never said it. Honest. Oh, I said there are maybe 100 billion galaxies and 10 billion trillion stars. It's hard to talk about the Cosmos without using big numbers. I said "billion" many times on the *Cosmos* television series, which was seen by a great many people. But I never said "billions and billions." For one thing, it's too imprecise. How many billions *are* "billions and billions"? A few billion? Twenty billion? A hundred billion? "Billions and billions" is pretty vague. When we reconfigured and updated the series, I checked — and sure enough, I never said it.

But Johnny Carson — on whose *Tonight Show* I'd appeared almost thirty times over the years — said it. He'd dress up in a corduroy jacket, a turtleneck sweater, and something like a mop for a wig. He had created a rough imitation of me, a kind of Doppelgänger, that went around saying "billions and billions" on late-night television. It used to bother me a little to have some simulacrum of my persona wandering off on its own, saying things that friends and colleagues would report to me the next morning. (Despite the disguise, Carson — a serious amateur astronomer — would often make my imitation talk real science.)

Astonishingly, "billions and billions" stuck. People liked the sound of it. Even today, I'm stopped on the street or on an airplane or at a party and asked, a little shyly, if I wouldn't — just for them — say "billions and billions."

"You know, I didn't actually say it," I tell them.

"It's okay," they reply. "Say it anyway."

I'm told that Sherlock Holmes never said, "Elementary, my dear Watson" (at least in the Arthur Conan Doyle books); Jimmy Cagney never said, "You dirty rat"; and Humphrey Bogart never said, "Play it again, Sam." But they might as well have, because these apocrypha have firmly insinuated themselves into popular culture.

I'm still quoted as uttering this simple-minded phrase in computer magazines ("As Carl Sagan would say, it takes billions and billions of bytes"), newspaper economics primers, discussions of players' salaries in professional sports, and the like.

For a while, out of childish pique, I wouldn't utter or write the phrase, even when asked to. But I've gotten over that. So, for the record, here goes:

"Billions and billions."

What makes "billions and billions" so popular? It used to be that "millions" was the byword for a large number. The enormously rich were millionaires. The population of the Earth at the time of Jesus was perhaps 250 million people. There were almost 4 million Americans at the time of the Constitutional Convention of 1787; by the beginning of World War II, there were 132 million. It is 93 million miles (150 million kilometers) from the Earth to the Sun. Approximately 40 million people were killed in World War I; 60 million in World War II. There are 31.7 million seconds in a year (as is easy enough to verify). The global nuclear arsenals at the end of the 1980s contained an equivalent explosive power sufficient to destroy 1 million Hiroshimas. For many purposes and for a long time, "million" was the quintessential big number.

But times have changed. Now the world has a clutch of *b*illionaires — and not just because of inflation. The age of the Earth is well-established at 4.6 billion years. The human population is pushing 6 billion people. Every birthday represents another billion kilometers around the Sun (the Earth is traveling around the Sun much faster than the *Voyager* spacecraft are traveling away from the Earth). Four B-2 bombers cost a billion dollars. (Some say 2 or even 4 billion.) The U.S. defense budget is, when hidden costs are accounted for, over $300 billion a year. The immediate fatalities in an all-out nuclear war between the United States and Russia are estimated to be around a billion people. A few inches are a billion atoms side by side. And there are all those billions of stars and galaxies.

In 1980, when the *Cosmos* television series

was first shown, people were ready for billions. Mere millions had become a little downscale, unfashionable, miserly. Actually, the two words sound sufficiently alike that you have to make a serious effort to distinguish them. This is why, in *Cosmos*, I pronounced "billions" with a fairly plosive "b," which some people took for an idiosyncratic accent or speech deficiency. The alternative, pioneered by TV commentators — to say, "That's billions with a *b*" — seemed more cumbersome.

There's an old joke about the planetarium lecturer who tells his audience that in 5 billion years the Sun will swell to become a bloated red giant, engulfing the planets Mercury and Venus and eventually perhaps even gobbling up the Earth. Afterward, an anxious member of the audience buttonholes him:

"Excuse me, Doctor, did you say that the Sun will burn up the Earth in 5 billion years?"

"Yes, more or less."

"Thank God. For a moment I thought you said 5 *m*illion."

Whether it's 5 million or 5 billion, it has little bearing on our personal lives, as interesting as the ultimate fate of the Earth may be. But the distinction between millions and billions is much more vital on such issues as national budgets, world population, and nuclear war fatalities.

While the popularity of "billions and billions" has not entirely faded, these numbers too are becoming somewhat small-scale, narrow-visioned, and passé. A much more fashionable number is now on the horizon, or closer. The *trillion* is almost upon us.

World military expenditures are now nearly $1 trillion a year. The total indebtedness of all developing nations to Western banks is pushing $2 trillion (up from $60 billion in 1970). The annual budget of the U.S. Government is also approaching $2 trillion. The national debt is around $5 trillion. The proposed, and technically dubious, Reagan-era Star Wars scheme was estimated to cost between $1 trillion and $2 trillion. All the plants on Earth weigh a trillion tons. Stars and trillions have a natural affinity: The distance from our Solar System to the nearest star, Alpha Centauri, is 25 trillion miles (about 40 trillion kilometers).

Confusion among millions, billions, and trillions is still endemic in everyday life, and it is a rare week that goes by without some such muddle on TV news (generally a mix-up between millions and billions). So, perhaps I can be excused for spending a moment distinguishing: A million is a thousand thousand, or a one followed by six zeros; a billion is a thousand million, or a one followed by nine zeros; and a trillion is a thousand billion (or equivalently, a million million), which is a one followed by 12 zeros.

This is the American convention. For a long time, the British word "billion" corresponded to the American "trillion," with the British using — sensibly enough — "thousand million" for a billion. In Europe, "milliard" was the word for a billion. As a stamp collector since childhood, I have an uncanceled postage stamp from the height of the 1923 German inflation that reads "50 milliarden." It took 50 trillion marks to mail a letter. (This was when people brought a wheelbarrow full of currency to the baker's or the grocer's.) But because of the current world influence of the United States, these alternative conventions are in retreat, and "milliard" has almost disappeared.

An unambiguous way to determine what large number is being discussed is simply to count up the zeros after the one. But if there are many zeros, this can get a little tedious. That's why we put commas, or spaces, after each group of three zeros. So a trillion is 1,000,000,000,000 or 1 000 000 000 000.

(In Europe one puts dots in place of commas.) For numbers bigger than a trillion, you have to count up how many triplets of 0s there are. It would be much easier still if, when we name a large number, we could just say straight out how many zeros there are after the one.

Scientists and mathematicians, being practical people, have done just this. It's called exponential notation. You write down the number 10; then a little number, written above and to the right of the 10 as a superscript, tells how many zeros there are after the one. Thus, 10^6 = 1,000,000; 10^9 = 1,000,000,000; 10^{12} = 1,000,000,000,000; and so on. These little superscripts are called exponents or powers; for example, 10^9 is described as "10 to the power 9" or, equivalently, "10 to the ninth" (except for 10^2 and 10^3, which are called "10 squared" and "10 cubed," respectively). This phrase, "to the power" — like "parameter" and a number of other scientific and mathematical terms — is creeping into everyday language, but with the meaning progressively blurred and distorted.

In addition to clarity, exponential notation has a wonderful side benefit: You can multiply any two numbers just by adding the appropriate exponents. Thus, 1,000 x

1,000,000,000 is $10^3 \times 10^9 = 10^{12}$. Or take some larger numbers: If there are 10^{11} stars in a typical galaxy and 10^{11} galaxies, there are 10^{22} stars in the Cosmos.

But there is still resistance to exponential notation from people a little jittery about mathematics (although it simplifies, not complicates, our understanding) and from typesetters, who seem to have a passionate need to print 10^9 as 109 (the typesetters for Random House and Thorndike Press being a welcome exception, as you can see).

The first six big numbers that have their own names are shown in the accompanying box. Each is 1,000 times bigger than the one before. Above a trillion, the names are almost never used. You could count one number every second, day and night, and it would take you more than a week to count from one to a million. A billion would take you half a lifetime. And you couldn't get to a quintillion even if you had the age of the Universe to do it in.

Once you've mastered exponential notation, you can deal effortlessly with immense numbers, such as the rough number of microbes in a teaspoon of soil (10^8); of grains of sand on all the beaches of the Earth (maybe 10^{20}); of living things on Earth (10^{29}); of atoms in all the life on

BIG NUMBERS

Name (U.S.)	Number (written out)	Number (scientific notation)	How long it would take to count to this number from 0 (one count per second, night and day)
One	1	10^0	1 second
Thousand	1,000	10^3	17 minutes
Million	1,000,000	10^6	12 days
Billion	1,000,000,000	10^9	32 years
Trillion	1,000,000,000,000	10^{12}	32,000 years (longer than there has been civilization on Earth)
Quadrillion	1,000,000,000,000,000	10^{15}	32 million years (longer than there have been humans on Earth)
Quintillion	1,000,000,000,000,000,000	10^{18}	32 billion years (more than the age of the Universe)

Larger numbers are called a sextillion (10^{21}), septillion (10^{24}), octillion (10^{27}), nonillion (10^{30}), and decillion (10^{33}). The Earth has a mass of 6 octillion grams.

That scientific or exponential notation also is described by words. Thus, an electron is a femtometer (10^{-15} m) across; yellow light has a wavelength of half a micrometer (0.5 μm); the human eye can barely see a bug a tenth of a millimeter (10^{-4} m) across; the Earth has a radius of 6,300 kilometers (6,300 km = 6.3 Mm); and a mountain might weigh 100 petagrams (100 pg = 10^{17} g). A complete list of prefixes goes as follows:

atto-	a	10^{-18}	deka-	—	10^{1}
femto-	f	10^{-15}	hecto-	—	10^{2}
pico-	p	10^{-12}	kilo-	k	10^{3}
nano-	n	10^{-9}	mega-	M	10^{6}
micro-	μ	10^{-6}	giga-	G	10^{9}
milli-	m	10^{-3}	tera-	T	10^{12}
centi-	c	10^{-2}	peta-	P	10^{15}
deci-	d	10^{-1}	exa-	E	10^{18}

1... 2... 3... 4....
231,254,321,974...231,254,321,975...

311,567,075,686.... 311,567,075,687....
..362,504,379,077....362,504,379,078..

..374,379,317,420....374,379,317,421...
..378,000,000,001...378,000,000,002.....

Earth (10^{41}); of atomic nuclei in the Sun (10^{57}); or of the number of elementary particles (electrons, protons, neutrons) in the entire Cosmos (10^{80}). This doesn't mean you can *picture* a billion or a quintillion objects in your head — nobody can. But, with exponential notation, we can *think* about and calculate with such numbers. Pretty good for self-taught beings who started out with no possessions and who could number their fellows on their fingers and toes.

Really big numbers are part and parcel of modern science; but I don't want to leave the impression that they were invented in our time.

Indian arithmetic has long been equal to large numbers. You can easily find references in Indian newspapers today to fines or expenditures of lakh or crore rupees. The key is: das = 10; san = 100; hazar = 1,000; lakh = 10^5; crore = 10^7; arahb = 10^9; carahb = 10^{11}; nie = 10^{13}; padham = 10^{15}; and sankh = 10^{17}. Before their culture was annihilated by the Europeans, the Maya of ancient Mexico devised a world timescale that dwarfed the paltry few thousand years that the Europeans thought had passed since the creation of the world. Among the decaying monuments of Coba, in Quintana

Roo, are inscriptions showing that the Maya were contemplating a Universe around 10^{29} years old. The Hindus held that the present incarnation of the Universe is 8.6×10^9 years old — almost right on the button. And the third century B.C. Sicilian mathematician Archimedes, in his book *The Sand-Reckoner*, estimated that it would take 10^{63} grains of sand to fill the Cosmos. On the really big questions, billions and billions were small change even then.

Chapter 2

THE PERSIAN CHESSBOARD

There cannot be a language more universal and more simple, more free from errors and from obscurities, that is to say more worthy to express the invariable relations of natural things. . . . [Mathematics] seems to be a faculty of the human mind destined to supplement the shortness of life and the imperfection of the senses.

JOSEPH FOURIER,
Analytic Theory of Heat,
Preliminary Discourse (1822)

The way I first heard the story, it happened in ancient Persia. But it may have been India or even China. Anyway, it happened a long time ago. The Grand Vizier, the principal advisor to the King, had invented a new game. It was played with moving pieces on a square board comprised of 64 red and black squares. The most important piece was the King. The next most important piece was the Grand Vizier — just what we might expect of a game invented by a Grand Vizier. The object of the game was to capture the enemy King, and so the game was called, in Persian, *shahmat* — *shah* for King, *mat* for dead. Death to the King. In Russian it is still called *shakhmat,* which perhaps conveys a lingering revolutionary sentiment. Even in English there is an echo of this

name — the final move is called "checkmate." The game, of course, is chess. As time passed, the pieces, their moves, and the rules of the game all evolved; there is, for example, no longer a Grand Vizier — it has become transmogrified into a Queen, with much more formidable powers.

Why a King should delight in the invention of a game called "Death to the King" is a mystery. But, so the story goes, he was so pleased that he asked the Grand Vizier to name his own reward for so splendid an invention. The Grand Vizier had his answer ready: He was a modest man, he told the Shah. He wished only for a modest reward. Gesturing to the eight columns and eight rows of squares on the board he had invented, he asked that he be given a single grain of wheat on the first square, twice that on the second square, twice *that* on the third, and so on, until each square had its complement of wheat. No, the King remonstrated, this is too modest a reward for so important an invention. He offered jewels, dancing girls, palaces. But the Grand Vizier, his eyes becomingly lowered, refused them all. It was little piles of wheat that he craved. So, secretly marveling at the humility and restraint of his counselor, the King consented.

When, however, the Master of the Royal Granary began to count out the grains, the King faced an unpleasant surprise. The number of grains starts out small enough: 1, 2, 4, 8, 16, 32, 64, 128, 256, 512, 1024 . . . but by the time the 64th square is approached, the number of grains becomes colossal, staggering. In fact, the number is (see box on page 41) nearly 18.5 quintillion. Perhaps the Grand Vizier was on a high-fiber diet.

How much does 18.5 quintillion grains of wheat weigh? If each grain is a millimeter in size, then all of the grains together would weigh around 75 billion metric tons, which far exceeds what could have been stored in the Shah's granaries. In fact, this is the equivalent of about 150 years of the world's *present* wheat production. An account of what happened next has not come down to us. Whether the King, in default, blaming himself for inattentiveness in his study of arithmetic, handed the kingdom over to the Vizier, or whether the latter experienced the tribulations of a new game called *viziermat*, we are not privileged to know.

The story of the Persian Chessboard may be just a fable. But the ancient Persians and Indians were brilliant pathfinders in mathematics, and understood the enormous numbers that result when you keep on doubling. Had chess been invented with 100 (10 x 10) squares instead of 64 (8 x 8), the resulting debt in grains of wheat would have weighed as much as the Earth. A sequence of numbers like this, where each number is a fixed multiple of the previous one, is called a geometric progression, and the process is called an exponential increase.

Exponentials show up in all sorts of important areas, unfamiliar and familiar — for

erosexuals of both sexes, who perhaps find prudence overwhelmed by passion and use unsafe sexual practices. Many of them will die, some will be lucky or naturally immune or abstemious, and they will be replaced by another most-at-risk group — perhaps the next generation of homosexual men. Eventually the exponential curve for all of us together is expected to flatten out, having killed many fewer than everybody on Earth. (Small comfort for its many victims and their loved ones.)

Exponentials are also the central idea behind the world population crisis. For most of the time humans have been on Earth the population was stable, with births and deaths almost perfectly in balance. This is called a "steady state." After the invention of agriculture — including the planting and harvesting of those grains of wheat the Grand Vizier was hankering for — the human population of this planet began increasing, entering an exponential phase, which is very far from a steady state. Right now the doubling time of the world population is about 40 years. Every 40 years there will be twice as many of us. As the English clergyman Thomas Malthus pointed out in 1798, a population increas-

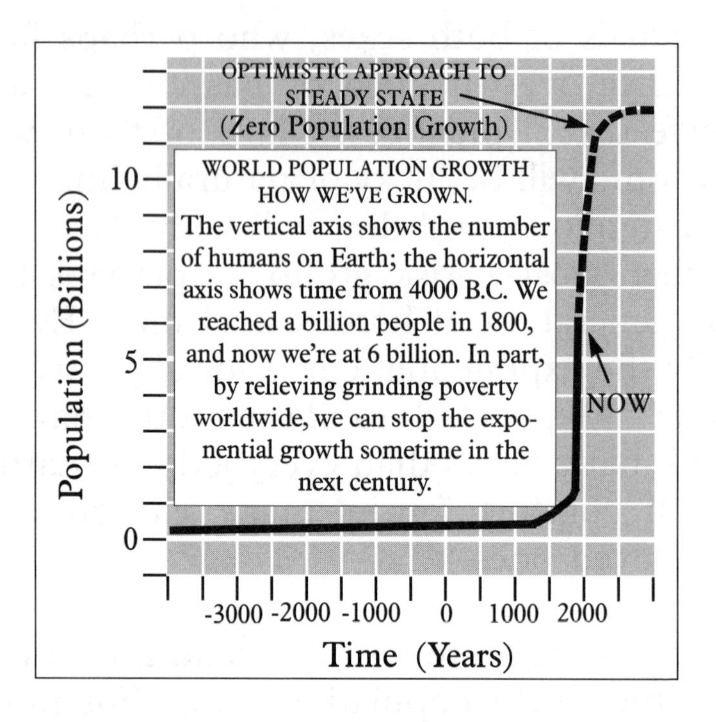

ing exponentially — Malthus described it as a geometrical progression — will outstrip any conceivable increase in food supply. No Green Revolution, no hydroponics, no making the deserts bloom can beat an exponential population growth.

There is also no extraterrestrial solution to this problem. Right now there are something like 240,000 more humans being born than dying every day. We are very far from being able to ship 240,000 people into space every day. No settlements in Earth orbit or on the Moon or on other planets can put a perceptible dent in the population explosion. Even if it were possible to ship everybody on Earth off to planets of distant stars on ships that travel faster than light, almost nothing would be changed — all the habitable planets in the Milky Way galaxy would be full up in a millennium or so. Unless we slow our rate of reproduction. Never underestimate an exponential.

The Earth's population, as it has grown through time, is shown in the previous figure. We are clearly in (or just about to emerge from) a phase of steep exponential growth. But many countries — the United States, Russia, and China, for example — have reached or will soon reach a situation where their population growth has ceased, where they arrive at something close to a steady state. This is also called zero population growth (ZPG). Still, because exponentials are so powerful, if even a small fraction of the human community continues for some time to reproduce exponentially the situation is essentially the same — the world population increases exponentially, even if many nations are at ZPG.

There is a well-documented worldwide correlation between poverty and high birthrates. In little countries and big countries, capitalist countries and communist countries, Catholic countries and Moslem countries, Western countries and Eastern countries — in almost all these cases, exponential population growth slows down or stops when grinding poverty disappears. This is called the demographic transition. It is in the urgent long-term interest of the human species that every place on Earth achieves this demographic transition. This is why helping other countries to become self-sufficient is not only elementary human decency, but is also in the self-interest of those richer nations able to help. One of the central issues in the world population crisis is poverty.

The exceptions to the demographic transition are interesting. Some nations with high per capita incomes still have high birthrates. But in them, contraceptives are sparsely available, and/or women lack any effective political power. It is not hard to understand the connection.

At present there are around 6 billion humans. In 40 years, if the doubling time stays constant, there will be 12 billion; in 80 years, 24 billion; in 120 years, 48 billion. . . . But few believe the Earth can support so many people. Because of the power of this exponential increase, dealing with global poverty now will be much cheaper and much more humane, it seems, than whatever solutions will be available to us many decades hence. Our job is to bring about a worldwide demographic transition and flatten out that exponential curve — by eliminating grinding poverty, making safe and effective birth control methods widely available, and extending real political power (executive, legislative, judicial, military, and in institutions influencing public opinion) to women. If we fail, some other process, less under our control, will do it for us.

Speaking of which . . .

Nuclear fission was first thought of in London in September 1933 by an émigré Hungarian physicist named Leo Szilard. He had been wondering whether human tinkering could unlock the vast energies hidden in the nucleus of the atom. He asked himself what would happen if a neutron were fired at an atomic nucleus. (Because it has no electrical charge, a neutron would not be electrically repelled by the protons in the nucleus, and would instead collide directly with the nucleus.) As he was waiting for a traffic signal to change at an intersection on Southhampton Row, it dawned on him that there might be some substance, some chemical element, which spat out two neutrons when it was hit by one. Each of *those* neutrons could eject more neutrons, and there suddenly appeared in Szilard's mind the vision of a nuclear chain reaction, with exponentiating numbers of neutrons produced and atoms falling to pieces left and right. That evening, in his small room in the Strand Palace Hotel, he calculated that only a few pounds of matter, if it could be made to undergo a controlled neutron chain reaction, might liberate enough energy to run a small city for a year . . . or, if the energy were released suddenly, enough to destroy that city utterly. Szilard eventually emigrated to the United States, and began a systematic search through all the chemical elements to see if any produced more neutrons than collided with them. Uranium seemed a promising candidate. Szilard convinced Albert Einstein to write his famous letter to President Roosevelt urging the United States to build an atomic bomb. Szilard played a major role in the first uranium chain reaction in Chicago in 1942, which in fact led to the atomic bomb. He spent the rest of his life warning about the dangers of the weapon he had been the first to conceive. He had found, in yet another way, the awesome power of the exponential.

Everybody has 2 parents, 4 grandparents, 8 great-grandparents, 16 great-great-grand-

THE CALCULATION THE KING SHOULD HAVE DEMANDED OF HIS VIZIER

Don't be scared off. This is really easy. We want to calculate how many grains of wheat were on the entire Persian Chessboard.

An elegant (and perfectly exact) calculation goes as follows:

The exponent just tells us how many times we multiply 2 by itself. $2^2 = 4$. $2^4 = 16$. $2^{10} = 1{,}024$, and so on. Call S the total number of grains on the chessboard, from 1 in the first square to 2^{63} in the 64th square. Then, plainly,

$$S = 1 + 2 + 2^2 + 2^3 + \ldots + 2^{62} + 2^{63}$$

Simply by doubling both sides of the last equation, we find

$$2S = 2 + 2^2 + 2^3 + 2^4 + \ldots + 2^{63} + 2^{64}$$

Subtracting the first equation from the second gives us

$$2S - S = S = 2^{64} - 1,$$

which is the exact answer.

How much is it roughly in ordinary base-10 notation? 2^{10} is close to 1,000, or 10^3 (within 2.4 percent). So $2^{20} = 2^{(10 \times 2)} = (2^{10})^2$ = roughly $(10^3)^2 = 10^6$, which is 10 multiplied by itself 6 times, or a million. Likewise, $2^{60} = (2^{10})^6$ = roughly $(10^3)^6 = 10^{18}$. So $2^{64} = 2^4 \times 2^{60}$ = roughly 16×10^{18}, or 16 followed by 18 zeros, which is 16 quintillion grains. A more accurate calculation gives the answer 18.6 quintillion grains.

parents, etc. Every generation back we go, we have twice as many lineal ancestors. You can see that this is very much a Persian Chessboard kind of problem. If there are, say, 25 years to a generation, then 64 generations ago is 64 x 25 = 1,600 years ago, or just before the fall of the Roman Empire. So (see box), every one of us alive today had in the year 400 some 18.5 quintillion ancestors — or so it seems. And this says nothing about collateral relatives. But this is far more than the population of the Earth, then or now; it is far more than the total number of human beings who have ever lived. Something is wrong with our calculation. What? Well, we have assumed all those lineal ancestors to be different people. But this, of course, is not the case. The same ancestor is related to us by many different routes. We are repeatedly, multiply connected with each of our relatives — a huge number of times for the more distant relations.

Something like this is true of the whole human population. If we go far enough back, any two people on Earth have a common ancestor. Whenever a new American President is elected, there's bound to be someone — generally in England — to discover that the new President is related to the Queen or King of England. This is thought to bind the English-speaking peoples together. When two people derive from the same nation or culture, or from the same small corner of the world, and their genealogies are well-recorded, it is likely that the last common ancestor can be discovered. But whether it can be discovered or not, the relationships are clear. We are all cousins — everyone on Earth.

Another common appearance of exponentials is the idea of half-life. A radioactive "parent" element — plutonium, say, or radium — decays into another, perhaps safer, "daughter" element, but not all at once. It decays statistically. There is a certain time by which half of it has decayed, and this is called its half-life. Half of what is left decays in another half-life, and half of the remainder in still another half-life, and so on. For example, if the half-life were one year, half would decay in a year, half of a half or all but a quarter would be gone in two years, all but an eighth in three years, all but about a thousandth in ten years, etc. Different elements have different half-lives. Half-life is an important idea when trying to decide what to do with radioactive waste from nuclear power plants or in contemplating radioactive fallout in nuclear war. It represents an exponential decay, in the same way that the Persian Chessboard represents an exponential increase.

Radioactive decay is a principal method for dating the past. If we can measure the amount of radioactive parent material and the amount of daughter decay product in a sample, we can determine how long the sample has been around. In this way we find that the so-called Shroud of Turin is not the burial shroud of Jesus, but a pious hoax from the fourteenth century (when it was denounced by Church authorities); that humans made campfires millions of years ago; that the most ancient fossils of life on Earth are at least 3.5 billion years old; and that the Earth itself is 4.6 billion years old. The Cosmos, of course, is billions of years older still. If you understand exponentials, the key to many of the secrets of the Universe is in your hand.

If you know a thing only qualitatively, you know it no more than vaguely. If you know it quantitatively — grasping some numerical measure that distinguishes it from an infinite number of other possibilities — you are beginning to know it deeply. You comprehend some of its beauty and you gain access to its power and the understanding it provides. Being afraid of quantification is tantamount to disenfranchising yourself, giving up on one of the most potent prospects for understanding and changing the world.

Chapter 3

MONDAY-NIGHT HUNTERS

The hunting instinct has [a] . . . remote origin in the evolution of the race. The hunting and the fighting instinct combine in many manifestations. . . . It is just because human bloodthirstiness is such a primitive part of us that it is so hard to eradicate, especially where a fight or a hunt is promised as part of the fun.

WILLIAM JAMES,
Psychology, XXIV (1890)

We can't help ourselves. On Sunday afternoons and Monday nights in the fall of each year, we abandon everything to watch small moving images of 22 men — running into one another, falling down, picking themselves up, and kicking an elongated object made from the skin of an animal. Every now and then, both the players and the sedentary spectators are moved to rapture or despair by the progress of the play. All over America, people (almost exclusively men), transfixed before glass screens, cheer or mutter in unison. Put this way, it sounds stupid. But once you get the hang of it, it's hard to resist, and I speak from experience.

Athletes run, jump, hit, slide, throw, kick, tackle — and there's a thrill in seeing humans do it so well. They wrestle each other to the ground. They're keen on grabbing or clubbing or kicking a fast-moving brown or white thing. In some games, they try to herd the thing toward what's called a "goal"; in other games, the players run away and then return "home." Teamwork is almost everything, and we admire how the parts fit together to make a jubilant whole.

But these are not the skills by which most of us earn our daily bread. Why should we feel compelled to watch people run or hit? Why is this need transcultural? (Ancient Egyptians, Persians, Greeks, Romans, Mayans, and Aztecs also played ball. Polo is Tibetan.)

There are sports stars who make 50 times the annual salary of the President; some who are themselves, after retirement, elected to high office. They are national heroes. Why, exactly? There is something here transcending the diversity of political, social, and economic systems. Something ancient is calling.

Most major sports are associated with a nation or a city, and they carry with them elements of patriotism and civic pride. Our team represents *us* — where we live, our people — against those other guys from some different place, populated by unfamiliar, maybe hostile people. (True, most of "our" players are not *really* from here. They're mercenaries and with clear conscience regularly defect from opposing cities for suitable emolument: A Pittsburgh Pirate is reformed into a California Angel; a San Diego Padre is raised to a St. Louis Cardinal; a Golden State Warrior is crowned a Sacramento King. Occasionally, a whole team picks up and migrates to another city.)

Competitive sports are symbolic conflicts, thinly disguised. This is hardly a new insight. The Cherokees called their ancient form of lacrosse "the little brother of war." Or here is Max Rafferty, former California Superintendent of Public Instruction, who, after denouncing critics of college football as "kooks, crumbums, commies, hairy loudmouthed beatniks," goes on to state, "Football players . . . possess a clear, bright, fighting spirit which is America itself." (That's worth mulling over.) An often-quoted sentiment of the late professional football coach Vince Lombardi is that the only thing that counts is winning. Former Washington Redskins' coach George Allen put it this way: "Losing is like death."

Indeed, we talk of winning and losing a war as naturally as we do of winning and losing a game. In a televised U.S. Army recruitment ad, we see the aftermath of an armored warfare exercise in which one tank destroys another; in the tag line, the victorious tank commander says, "When we win, the whole team wins — not one person." The connection between sports and combat is made quite clear. Sports fans (the word is short for "fanatics") have been known to commit assault and battery, and sometimes murder, when taunted about a losing team; or when prevented from cheering on a winning team; or when they feel an injustice has been committed by the referees.

The British Prime Minister was obliged in 1985 to denounce the rowdy, drunken behavior of British soccer fans who attacked an Italian contingent for having the effrontery to root for their own team. Dozens were killed when the stands collapsed. In 1969, after three hard-fought soccer games, Salvadoran tanks crossed the Honduran border, and Salvadoran bombers attacked Honduran ports and military bases. In this "Soccer War," the casualties numbered in the thousands.

Afghan tribesmen played polo with the severed heads of former adversaries. And 600 years ago, in what is now Mexico City, there was a ball court where gorgeously attired nobles watched uniformed teams compete. The captain of the losing team was beheaded, and the skulls of earlier losing captains were displayed on racks — an inducement possibly even more compelling than winning one for the Gipper.

Suppose you're idly flipping the dial on your television set, and you come upon some competition in which you have no particular emotional investment — say, off-season volleyball between Myanmar and Thailand. How do you decide which team to root for? But wait a minute: Why root for either? Why not just enjoy the game? Most of us have trouble with this detached posture. We want to take part in the contest, to feel ourselves a member of a team. The feeling simply sweeps us away, and there we are rooting, "Go, Myanmar!" Initially, our loyalties may oscillate, first urging on one team and then the other. Sometimes we root for the underdog. Other times, shamefully, we even switch our allegiance from loser to winner as the outcome becomes clear. (When there is a succession of losing seasons, fan loyalties tend to drift elsewhere.) What we are looking for is victory without effort. We want to be swept up into something like a small, safe, successful war.

In 1996, Mahmoud Abdul-Rauf, then a guard for the Denver Nuggets, was suspended by the National Basketball Association. Why? Because Abdul-Rauf refused to stand for the compulsory playing of the National Anthem. The American flag represented to him a "symbol of oppression" offensive to his Muslim beliefs. Most other players, while not sharing Abdul-Rauf's beliefs, supported his right to express them. Harvey Araton, a distinguished sports writer for the *New York Times*, was puzzled. Playing the anthem at a sporting event "is, let's face it, a tradition that is absolutely idiotic in today's world," he explains, "as opposed to when it began, before baseball games during World War II. Nobody goes to a sporting event to make an expression of patriotism." On the contrary, I would argue that a kind of patriotism and nationalism is very much what sporting events are about.*

The earliest known organized athletic events date back 3,500 years to preclassical Greece. During the original Olympic Games, an armistice put all wars among Greek city-states on hold. The games were more important than the wars. The men performed nude: No women spectators were allowed. By the eighth century B.C., the Olympic Games consisted of running (*lots* of running), jumping, throwing things (including javelins), and wrestling (sometimes to the death). While none of these events was a team sport, they are clearly central to modern team sports.

They were also central to low-technology hunting. Hunting is traditionally considered a sport, as long as you don't eat what you catch — a proviso much easier for the rich to comply with than the poor. From the earliest pharaohs, hunting has been associated with military aristocracies. Oscar Wilde's aphorism about English fox hunting, "the unspeakable in full pursuit of the uneatable," makes a similar dual point. The forerunners of football, soccer, hockey, and kin-

*The crisis was resolved when Mr. Abdul-Rauf agreed to stand during the anthem, but pray instead of sing.

dred sports were disdainfully called "rabble games," recognized as substitutes for hunting — because young men who worked for a living were barred from the hunt.

The weapons of the earliest wars must have been hunting implements. Team sports are not just stylized echoes of ancient wars. They also satisfy an almost-forgotten craving for the hunt. Since our passions for sports run so deep and are so broadly distributed, they are likely to be hardwired into us — not in our brains but in our genes. The 10,000 years since the invention of agriculture is not nearly enough time for such predispositions to have evolved away and disappeared. If we want to understand them, we must go much further back.

The human species is hundreds of thousands of years old (the human family several millions of years old). We have led a sedentary existence — based on farming and domestication of animals — for only the last 3 percent of that period, during which is all our recorded history. In the first 97 percent of our tenure on Earth, almost everything that is characteristically human came into being. So a little arithmetic about our history suggests we can learn something about those times from the few surviving hunter-gatherer communities uncorrupted by civilization.

* * *

We wander. With our little ones and all our belongings on our backs, we wander — following the game, seeking the water holes. We set up camp for a time, then move on. In providing food for the group, the men mainly hunt, the women mainly gather. Meat and potatoes. A typical itinerant band, mainly an extended family of relatives and in-laws, numbers a few dozen; although annually many hundreds of us, with the same language and culture, gather — for religious ceremonies, to trade, to arrange marriages, to tell stories. There are many stories about the hunt.

I'm focusing here on the hunters, who are men. But the women have significant social, economic, and cultural power. They gather the essential staples — nuts, fruits, tubers, roots — as well as medicinal herbs, hunt small animals, and provide strategic intelligence on large animal movements. Men do some gathering as well, and considerable "housework" (even though there are no houses). But hunting — only for food, never for sport — is the lifelong occupation of every able-bodied male.

Preadolescent boys stalk birds and small mammals with bows and arrows. By adulthood they have become experts in weapons procurement; in stalking, killing, and butchering the prey; and in carrying the cuts of meat back to camp. The first successful kill of a large mammal marks a young man's coming of age. In his initiation, ceremonial incisions are made on his chest or arms and an herb is rubbed into the cuts so that, when healed, a patterned tattoo results. It's like campaign ribbons — one look at his chest, and you know something of his combat experience.

From a jumble of hoofprints, we can accurately tell how many animals passed; the species, sexes, and ages; whether any are lame; how long ago they passed; how far away they are. Some young animals can be caught by open-field tackles; others with slingshots or boomerangs, or just by throwing rocks accurately and hard. Animals that have not yet learned to fear men can be approached boldly and clubbed to death. At greater distances, for warier prey, we hurl spears or shoot poisoned arrows. Sometimes we're lucky and, by a skillful rush, drive a herd of animals into an ambush or off a cliff.

Teamwork among the hunters is essential. If we are not to frighten the quarry, we must communicate by sign language. For the same reason, we need to have our emotions under control; both fear and exultation are dangerous. We are ambivalent about the prey. We respect the animals, recognize our kinship, identify with them. But if we reflect too closely on their intelligence or devotion to their young, if we feel pity for them, if we too deeply recognize them as relatives, our dedication to the hunt will slacken; we will bring home less food, and again our band may be endangered. We are obliged to put an emotional distance between us and them.

So contemplate this: For millions of years, our male ancestors are scampering about, throwing rocks at pigeons, running after baby antelopes and wrestling them to the ground, forming a single line of shouting, running hunters and trying to terrify a herd of startled warthogs upwind. Imagine that their lives depend on hunting skills and teamwork. Much of their culture is woven on the loom of the hunt. Good hunters are also good warriors. Then, after a long while — a few thousand centuries, say — a natural predisposition for both hunting and teamwork will inhabit many newborn boys. Why? Because incompetent or unenthusiastic hunters leave fewer offspring. I don't think how to chip a spearpoint out of stone or how to feather an arrow is in our genes. That's taught or figured out. But a zest for the chase — I bet that *is* hardwired. Natural selection helped mold our ancestors into superb hunters.

The clearest evidence of the success of the hunter-gatherer lifestyle is the simple fact that it extended to six continents and lasted millions of years (to say nothing of the hunting proclivities of nonhuman primates). Those big numbers speak profoundly. After 10,000 generations in which the killing of animals was our hedge against starvation, those inclinations must still be in us. We hunger to put them to use, even vicariously. Team sports provide one way.

Some part of our beings longs to join a small band of brothers on a daring and intrepid quest. We can even see this in role-playing and computer games popular with prepubescent and adolescent boys. The traditional manly virtues — taciturnity, resourcefulness, modesty, accuracy, consistency, deep knowledge of animals, teamwork, love of the outdoors — were all adaptive behavior in hunter-gatherer times. We still admire these traits, although we've almost forgotten why.

Besides sports, there are few outlets available. In our adolescent males, we can still recognize the young hunter, the aspirant warrior — leaping across apartment roof-tops; riding, helmetless, on a motorcycle; making trouble for the winning team at a postgame celebration. In the absence of a steadying hand, those old instincts may go a little askew (although our murder rate is about the same as among the surviving hunter-gatherers). We try to ensure that any residual zest for killing does not spill over onto humans. We don't always succeed.

I think of how powerful those hunting instincts are, and I worry. I worry that Monday-night football is insufficient outlet for the modern hunter, decked out in his overalls or jeans or three-piece suit. I think of that ancient legacy about not expressing our feelings, about keeping an emotional distance from those we kill, and it takes some of the fun out of the game.

Hunter-gatherers generally posed no danger to themselves: because their economies tended to be healthy (many had more free time than we do); because, as nomads, they had few possessions, almost no theft, and little envy; because greed and arrogance were considered not only social evils but also pretty close to mental illnesses; because women had real political power and tended to be a stabilizing and mitigating influence before the boys started going for their poisoned arrows; and because, when serious crimes were committed —

murder, say — the band collectively rendered judgment and punishment. Many hunter-gatherers organized egalitarian democracies. They had no chiefs. There was no political or corporate hierarchy to dream of climbing. There was no one to revolt against.

So, if we're stranded a few hundred centuries from when we long to be — if (through no fault of our own) we find ourselves, in an age of environmental pollution, social hierarchy, economic inequality, nuclear weapons, and declining prospects, with Pleistocene emotions but without Pleistocene social safeguards — perhaps we can be excused for a little Monday-night football.

TEAMS AND TOTEMS

Teams associated with cities have names: the Seibu Lions, the Detroit Tigers, the Chicago Bears. Lions and tigers and bears . . . eagles and seahawks . . . flames and suns. Allowing for the difference in environment and culture, hunter-gatherer groups worldwide have similar names — sometimes called totems.

A typical list of totems, mainly from the era before European contact, was recorded by the anthropologist Richard Lee in his many years among the !Kung "Bushmen" of the Kalahari Desert in Botswana (see page 65). The Short Feet, I think, are cousins to the Red Sox and White Sox, the Fighters to the Raiders, the Wildcats to the Bengals, the Cutters to the Clippers. Of course there are differences — due to technological differences and, perhaps, to varying endowments of candor, self-knowledge, and sense of humor. It's hard to imagine an American sports team named the Diarrheas ("Gimme a 'D' . . . "). Or — my personal favorite, a group of men with no self-esteem problems — the Big Talkers. And one in which the players are called the Owners would probably cause some consternation in the front office.

"Totemic" names are listed, top to bottom, in the following categories: birds, fish, mammals, and other animals; plants and minerals; technology; people, clothing, and occupations; mythical, religious, astronomical, and geological allusions; colors.

NORTH AMERICAN N.B.A. BASKETBALL

Hawks
Raptors
Bucks
Bulls
Grizzlies
Timberwolves
Hornets
Nuggets
Clippers
Heat
Pistons
Rockets
Spurs
Supersonics
Cavaliers
Celtics
Kings
Knickerbockers
Mavericks
Lakers
Nets
Pacers
76ers
Trail Blazers
Warriors

Jazz
Magic
Suns
Wizards

U.S.A. N.F.L. FOOTBALL

Cardinals
Eagles
Falcons
Ravens
Seahawks
Dolphins
Bears
Bengals
Bills
Broncos
Colts
Jaguars
Lions
Panthers
Rams
Jets
Buccaneers
Chargers
Chiefs
Cowboys
49ers
Oilers
Packers
Patriots

Raiders
Redskins
Saints
Steelers
Vikings
Giants
Browns

JAPANESE MAJOR LEAGUE BASEBALL

Hawks
Swallows
Carp
Buffaloes
Lions
Tigers
Whales
BayStars
Braves
Ham Fighters
Marines
Dragons
Giants
Orions
Blue Wave

NORTH AMERICAN MAJOR LEAGUE BASEBALL

Blue Jays
Cardinals

Orioles
Devil Rays
Marlins
Cubs
Tigers
Diamondbacks
Expos
Braves
Brewers
Dodgers
Indians
Twins
Yankees
Red Sox
White Sox
Athletics
Mets
Royals
Phillies
Pirates
Mariners
Rangers
Giants
Angels
Padres
Astros
Rockies
Reds

!KUNG NAMES GROUPS

Ant Bears
Elephants
Giraffes
Impalas
Jackals
Rhinos
Steenboks
Wildcats
Ants
Lice
Scorpions
Tortoises
Bitter Melons
Long Roots
Medicine Roots
Carrying Yokes
Cutters
Big Talkers
Cold Ones
Diarrheas
Dirty Fighters
Fighters
Owners
Penises
Short Feet

Chapter 4

THE GAZE OF GOD AND THE DRIPPING FAUCET

When you are risen on the eastern
  horizon
You have filled every land
  with your beauty . . .
Though you are far away, your rays are
  on Earth.

AKHNATON,
Hymn to the Sun (ca. 1370 B.C.)

In Pharaonic Egypt at the time of Akhnaton, in a now-extinct monotheistic religion that worshiped the Sun, light was thought to be the gaze of God. Back then, vision was imagined as a kind of emanation that proceeded *from* the eye. Sight was something like radar. It reached out and touched the object being seen. The Sun — without which little more than the stars are visible — was stroking, illuminating, and warming the valley of the Nile. Given the physics of the time, and a generation that worshiped the Sun, it made some sense to describe light as the gaze of God. Thirty-three hundred years later, a deeper, although much more prosaic metaphor provides a better understanding of light:

You're sitting in the bathtub, and the faucet is dripping. Once every second, say, a drop falls into the tub. It generates a little wave that spreads out in a beautiful perfect circle. As it reaches the sides of the tub, it's reflected back. The reflected wave is weaker, and after one or two more reflections, you can't make it out anymore.

New waves are arriving at your end of the tub, each generated by another drip of the faucet. Your rubber duck bobs up and down as each new wave front arrives before it. Clearly, the water is a little higher at the crest of the moving wave, and lower in the little shallow between the waves, the trough.

The "frequency" of the waves is simply how often the crests pass your vantage point — in this case, one wave every second. Since every drip makes a wave, the frequency is the same as the drip rate. The "wavelength" of the waves is simply the distance between successive wave crests — in this case, maybe 10 centimeters (about four inches). But if a wave passes every second, and they're ten centimeters apart, the speed of the waves is ten centimeters per second. The speed of a wave, you conclude after thinking about it a moment, is the frequency times the wavelength.

Bathtub waves and ocean waves are two-dimensional; they spread out from a point source as circles on the surface of the water. Sound waves, by contrast, are three-dimensional, spreading out in the air in all directions from the source of the sound. In the wave crest, the air is compressed a little; in the trough, the air is rarefied a little. Your ear detects these waves. The more often they come (the higher the frequency), the higher the pitch you hear.

Musical tones are only a matter of how often the sound waves strike your ears. Middle C is how we describe 263 sound waves reaching us every second; 263 hertz, it's called.* What would be the wavelength of Middle C? If sound waves were directly visible, how far would it be from crest to crest? At sea level, sound travels at about 340 meters per second (about 700 miles per hour). Just as in the bathtub, the wavelength will

*And one octave above Middle C is 526 hertz; two octaves, 1052 hertz; and so on.

be the speed of the wave divided by its frequency, or about 1.3 meters for Middle C — roughly, the height of a nine-year-old human.

There is a class of puzzle thought to confound science — which goes something like, "What is Middle C to a person deaf from birth?" Well, it's the same as it is to the rest of us: 263 hertz, a precise, unique frequency of sound belonging to this note and no other. If you can't hear it directly, you can detect it unambiguously with an audio amplifier and an oscilloscope. Now of course this isn't the same as experiencing the usual human perception of air waves — it utilizes sight rather than sound — but so what? All the information is there. You can sense chords and staccato, pizzicato, and timbre.

You can associate with other times you've "heard" Middle C. Maybe the electronic representation of Middle C isn't emotively the same as what a hearing person experiences, but even that may be a matter of experience. Even putting geniuses like Beethoven aside, you can be stone-deaf and experience music.

This is also the solution to the old conundrum about whether, if a tree falls in the forest and there's no one to hear, is a sound produced? Of course if we define a sound in terms of someone hearing it, by definition there was no sound. But this is an excessively anthropocentric definition. Clearly, if the tree falls, it makes sound waves, those sound waves can readily be detected by, say, a CD recorder, and when played back, the sound would be recognizably a tree falling in a forest. There is no mystery here.

But the human ear is not a perfect detector of sound waves. There are frequencies (fewer than 20 waves arriving per second) that are too low for us to hear, although whales communicate readily in such low tones. Likewise, there are frequencies (more than 20,000 waves arriving every second) too high-pitched for adult humans to detect, although dogs have no difficulty (and respond when called at such frequencies by a whistle). Realms of sound exist — a million waves per second, say — that are, and always will be, unknown to direct human perception. Our sense organs, as superbly adapted as they are, have fundamental physical limitations.

It's natural that we should communicate by sound. Our primate relatives certainly do. We're gregarious and mutually interdependent — there's a real necessity behind our communication talents. So, as our brains grew at an unprecedented rate over the last few million years, and as specialized regions of the cerebral cortex in charge of language evolved, our vocabulary proliferated. There was more and more that we were able to put into sounds.

When we were hunter-gatherers, language became essential for planning the day's activity, teaching the children, cementing friendships, alerting the others to danger, and sitting around the fire after dinner watching the stars come out and telling stories. Eventually, we invented phonetic writing so we could put our sounds down on paper and, by glancing at a page, hear someone speaking in our head — an invention that became so widespread in the last few thousand years that we hardly ever stop to consider how astonishing it is.

Speech is not really communicated instantaneously: When we make a sound, we are creating traveling waves in the air carried at the speed of sound. For practical purposes that's nearly instantaneous. But the trouble is that your shout carries only so far. It's a very rare person who can carry on a coherent conversation with someone even 100 meters away.

Until comparatively recently human population densities were very low. There was hardly any reason to communicate with someone more than 100 meters away. Almost no one — except members of our itinerant family group — ever came close enough to communicate with us. On the rare occasions that someone did, we were generally hostile. Ethnocentrism — the idea that our little group, no matter which one it is, is better than any other — and xenophobia — a "shoot first, ask questions later" fear of strangers — are deeply built into us. They are by no means peculiarly human; all our monkey and ape cousins behave similarly, as do many other mammals. These attitudes are at least aided and abetted by the short distances over which speech is possible.

If we're isolated for long periods from those other guys, we and they slowly develop in different directions. Their warriors start wearing ocelot skins, for example, instead of eagle feather headdresses — which everybody around here knows are fashionable, proper, and sane. Their language eventually becomes different from ours, their gods have strange names and demand bizarre ceremonies and sacrifices. Isolation breeds diversity; and our small numbers and limited communications range guarantee isolation. The human family — originating in one small locale in East Africa a few million years ago — wandered, separated, diversified, and became strangers to one another.

The reversal of this trend — the movement toward the reacquaintance and reunification of the lost tribes of the human family, the binding up of the species — has occurred only fairly recently and only because of advances in technology. The domestication of the horse permitted us to send messages (and ourselves) over distances of hundreds of miles in a few days. Advances in sailing ship technology allowed us to travel to the most distant reaches of the planet — but slowly: In the eighteenth century, it took about two years to sail from

Europe to China. By this time, far-flung human communities could send ambassadors to each other's courts, and exchange products of economic importance. However, for the great majority of eighteenth-century Chinese, Europeans could not have been more exotic had they lived on the Moon, and vice versa. The real binding up and deprovincialization of the planet requires a technology that communicates much faster than horse or sailing ship, that conveys information all over the world, and that is cheap enough to be available, at least occasionally, to the average person. Such a technology began with the invention of the telegraph and the laying of submarine cables; was greatly expanded by the invention of the telephone, using the same cables; and then enormously proliferated with the invention of radio, television, and satellite communications technology.

Today we communicate — routinely, casually, with hardly ever a second thought — at the speed of light. From the speed of horse or sailing ship to the speed of light is an improvement by a factor of almost a hundred million. For fundamental reasons at the heart of the way the world works, codified in Einstein's special theory of relativity, we know that there is no way we can send information faster than light. In a century, we have reached the ultimate speed limit. The technology is so powerful, its implications so far-reaching, that of course our societies have not yet caught up.

We place an overseas call, and we can sense that brief interval between when we finish asking a question and when the person we're talking to begins to answer. That delay is the time it takes for the sound our voice makes to get into the telephone, run electrically along the wires, reach a transmission station, be beamed up by microwaves to a communications satellite in geosynchronous orbit, be beamed down to a satellite receiving station, run through the wires some more, wiggle a diaphragm in a handset (halfway around the world, it may be), make sound waves in a very short length of air, enter someone's ear, carry an electrochemical message from ear to brain, and be understood.

The round-trip light travel time from the Earth to geosynchronous altitude is a quarter of a second. The farther apart the transmitter and receiver are, the longer it takes. In conversations with the *Apollo* astronauts on the Moon, the time delay between question and answer was longer. That was because the round-trip light (or radio) travel time between the Earth and the Moon is 2.6 seconds. It takes 20 minutes to receive a message from a spacecraft favorably situated in Martian orbit. In August 1989, we received pictures, taken by the *Voyager 2* spacecraft, of Neptune and its moons and ring arcs — data sent to us from the planetary frontiers of the Solar System, taking five hours to reach us at the speed of light. It was one of the longest long-distance calls ever placed by the human species.

In many contexts, light behaves as a wave. For example, imagine light passing through two parallel slits in a darkened room. What image does it cast on a screen behind the slits? Answer: an image of the slits — more exactly, a series of parallel bright and dark images of the slits — an "interference pattern." Rather than traveling like a bullet in a straight line, the waves spread from the two slits at various angles. Where crest falls on crest, we have a bright image of the slit: "constructive" interference; and where crest falls on trough, we have darkness: "destructive" interference. This is the signature behavior of a wave. You'd see the same thing with water waves and two holes cut at surface level in the pilings of a pier on a waterfront.

And yet light *also* behaves as a stream of

little bullets, called photons. This is how an ordinary photocell (in a camera, for instance, or a light-powered calculator) works. Each arriving photon ejects an electron from a sensitive surface; many photons generate many electrons, a flow of electric current. How can light simultaneously be a wave and a particle? It might be better to think of it as something else, neither a wave nor a particle, something with no ready counterpart in the everyday world of the palpable, that under some circumstances partakes of the properties of a wave, and, under others, of a particle. This wave-particle dualism is another reminder of a central humbling fact: Nature does not always conform to our predispositions and preferences, to what we deem comfortable and easy to understand.

And yet for most purposes, light is similar to sound. Light waves are three-dimensional, have a frequency, a wavelength, and a speed (the speed of light). But, astonishingly, they do not require a medium, like water or air, to propagate in. Light reaches us from the Sun and the distant stars, even though the intervening space is a nearly perfect vacuum. In space, astronauts without a radio link cannot hear each other, even if they are a few centimeters apart. There is no air to carry the sound.

But they can see one another perfectly well. Have them lean forward so their helmets touch, and they *can* hear one another. Take away all the air in your room and you will be unable to hear an acquaintance complain about it, although you will for a moment have no difficulty seeing him flailing and gasping.

For ordinary visible light — the kind our eyes are sensitive to — the frequency is very high, about 600 trillion (6×10^{14}) waves striking your eyeballs every second. Because the speed of light is 30 billion (3×10^{10}) centimeters a second (186,000 miles per second), the wavelength of visible light is about 30 billion divided by 600 trillion, or 0.00005 ($3 \times 10^{10}/6 \times 10^{14} = 0.5 \times 10^{-4}$) centimeters — much too small for us to see were it possible somehow for the waves themselves to be illuminated.

As different frequencies of sound are perceived by humans as different musical tones, so different frequencies of light are perceived as different colors. Red light has a frequency of about 460 trillion (4.6×10^{12}) waves per second, violet light about 710 trillion (7.1×10^{12}) waves per second. Between them are the familiar colors of the rainbow. Every color corresponds to a frequency.

As with the question of the meaning of a

musical tone to a person deaf since birth, there's the complementary question of the meaning of color to a person blind since birth. Again, the answer is uniquely and unambiguously a wave frequency — which can be measured optically and detected, if we so wish, as a musical tone. A blind person, properly trained and equipped in physics, can distinguish rose red from apple red from blood red. With the right kind of spectrometric library, she might be able to make much better compositional distinctions than the untrained human eye. Yes, there's a feeling of redness that sighted people sense around 460 trillion hertz. But I don't think that's anything more than what it feels like to sense 460 trillion hertz. There's no magic to it, as beautiful as it may be.

Just as there are sounds too high-pitched and too low-pitched for us to hear, so there are frequencies of light, or colors, outside our range of vision. They extend to much higher frequencies (around a billion billion* — 10^{18} — waves per second for gamma rays) and to much lower ones (less than one wave per second for long radio waves). Running through the spectrum of light from

*I know, I know. I can't help it: that's how many there are.

high frequency to low are broad swaths called gamma rays, X rays, ultraviolet light, visible light, infrared light, and radio waves. These are all waves that travel through a vacuum. Each is as legitimate a kind of light as ordinary visible light is.

There is an astronomy for each of these frequency ranges. The sky looks quite different in each regime of light. For example, bright stars are invisible in the light of gamma rays. But the enigmatic gamma ray bursters, detected by orbiting gamma ray observatories, are, so far, almost wholly indetectable in ordinary visible light. If we viewed the Universe in visible light only — as we did for most of our history — we would not know of the existence of gamma ray sources in the sky. The same is true of X-ray, ultraviolet, infrared, and radio sources (as well as the more exotic neutrino and cosmic ray sources, and — perhaps — gravity wave sources.)

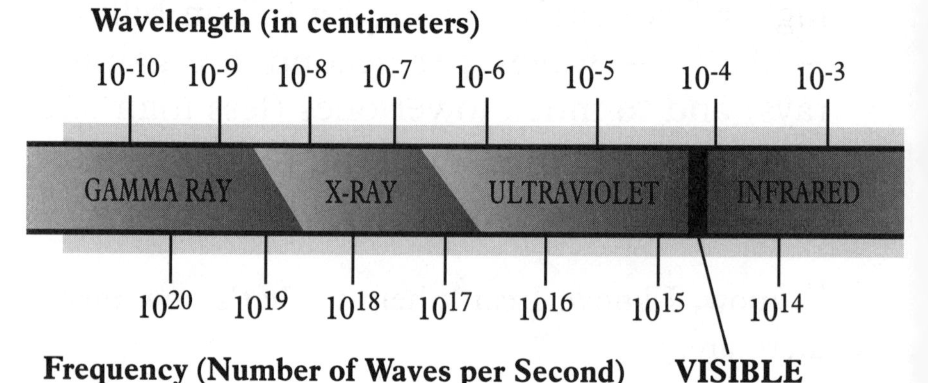

We're prejudiced toward visible light. We're visible light chauvinists. That's the only kind of light to which our eyes are sensitive. But if our bodies could transmit and receive radio waves, early humans might have been able to communicate with each other over great distances; if X rays, our ancestors might have peered usefully into the hidden interiors of plants, people, other animals, and minerals. So why didn't we evolve eyes sensitive to these other frequencies of light?

Any material you choose likes to absorb light of certain frequencies, but not of others. A different substance has a different preference. There is a natural resonance between light and chemistry. Some frequencies, such as gamma rays, are indiscriminately gobbled up by virtually all materials. If you had a gamma ray flashlight, the light would be readily absorbed by the air along its path. Gamma rays from space, traversing

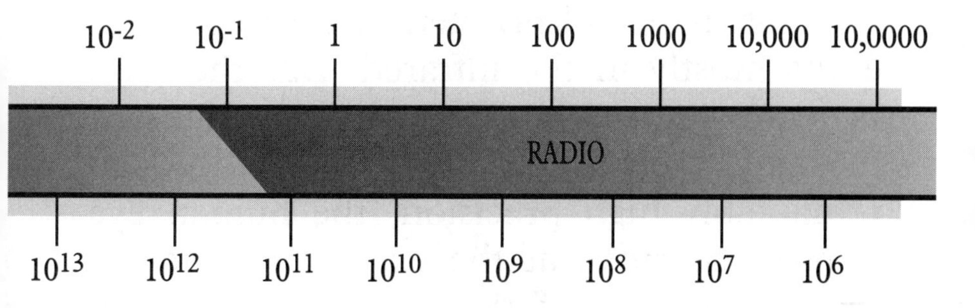

a much longer path through the Earth's atmosphere, would be entirely absorbed before they reached the ground. Down here on Earth, it's very dark in gamma rays — except around such things as nuclear weapons. If you want to see gamma rays from the center of the Galaxy, you must move your instruments into space. Something similar is true for X rays, ultraviolet light, and most infrared frequencies.

On the other hand, most materials are poor absorbers of visible light. Air, for example, is generally transparent to visible light. So one reason we see at visible frequencies is that this is the kind of light that gets through our atmosphere down to where we are. Gamma ray eyes would be of limited use in an atmosphere which makes things pitch black in gamma rays. Natural selection knows better.

The other reason we see in visible light is because that's where the Sun puts out most of its energy. A very hot star emits much of its light in the ultraviolet. A very cool star emits mostly in the infrared. But the Sun, in some respects an average star, puts out most of its energy in the visible. Indeed, to remarkably high precision, the human eye is most sensitive at the exact frequency in the yellow part of the spectrum at which the Sun is brightest.

Might the beings of some other planet see mainly at very different frequencies? This seems to me not at all likely. Virtually all cosmically abundant gases tend to be transparent in the visible and opaque at nearby frequencies. All but the coolest stars put out much, if not most, of their energy at visible frequencies. It seems to be only a coincidence that the transparency of matter and the luminosity of stars both prefer the same narrow range of frequencies. That coincidence applies not just to our Solar System, but throughout the Universe. It follows from fundamental laws of radiation, quantum mechanics, and nuclear physics. There might be occasional exceptions, but I think the beings of other worlds, if any, will probably see at very much the same frequencies as we do.*

Vegetation absorbs red and blue light, reflects green light, and so appears green to

*I still worry that some kind of visible light chauvinism plagues this argument: Beings like us who see only in visible light deduce that everyone in the entire Universe must see in visible light. Knowing how our history is rife with chauvinisms, I can't help being suspicious of my conclusion. But as nearly as I can see, it follows from physical law, not human conceit.

us. We could draw a picture of how much light is reflected at different colors. Something that absorbs blue and reflects red light appears to us red; something that absorbs red light and reflects blue appears to us blue. We see an object as white when it reflects light roughly equally in different colors. But this is also true of gray materials and black materials. The difference between black and white is not a matter of color, but of how much light they reflect. The terms are relative, not absolute.

Perhaps the brightest natural material is freshly fallen snow. But it reflects only about 75 percent of the sunlight falling on it. The darkest material that we ordinarily come into contact with — black velvet, say — reflects only a few percent of the light that falls on it. "As different as black and white" is a conceptual error: Black and white are fundamentally the same thing; the difference is only in the relative amounts of light reflected, not in their color.

Among humans, most "whites" are not as white as freshly fallen snow (or even a white refrigerator); most "blacks" are not as black as black velvet. The terms are relative, vague, confusing. The fraction of incident light that human skin reflects (the reflectivity) varies widely from individual to individ-

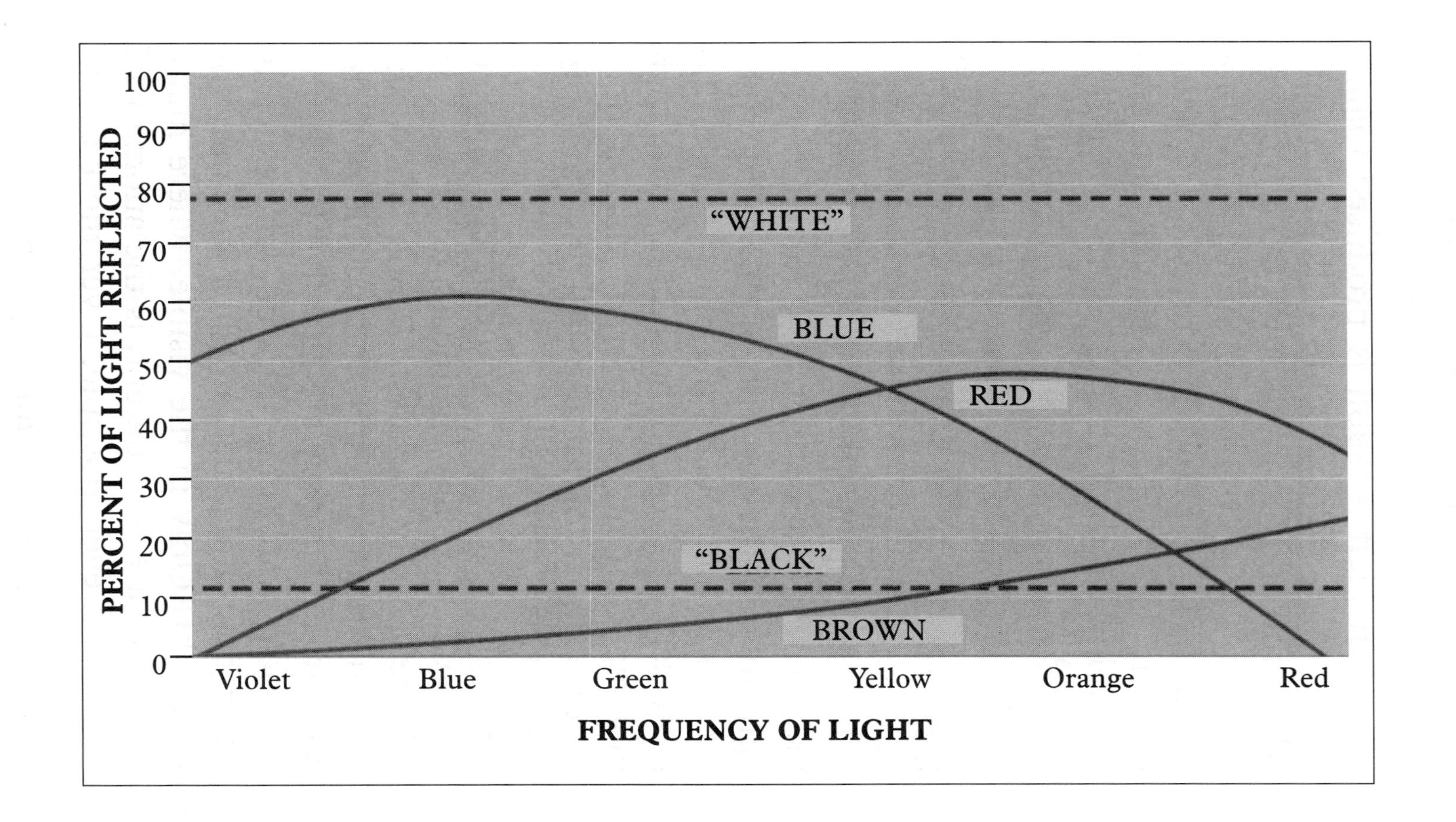
PERCENT OF LIGHT REFLECTED
100
90
80
70
60
50
40
30
20
10
0
"WHITE"
BLUE
RED
"BLACK"
BROWN
Violet
Blue
Green
Yellow
Orange
Red
FREQUENCY OF LIGHT

ual. Skin pigmentation is produced mainly by an organic molecule called melanin, which the body manufactures from tyrosine, an amino acid common in proteins. Albinos suffer from a hereditary disease in which melanin is not made. Their skin and hair are milky white. The irises of their eyes are pink. Albino animals are rare in Nature because their skins provide little protection against solar radiation, and because they lack protective camouflage. Albinos tend not to last long.

In the United States, almost everyone is brown. Our skins reflect somewhat more light toward the red end of the visible light spectrum than toward the blue. It makes no more sense to describe individuals with high melanin content as "colored" than it does to describe individuals with low melanin content as "bleached."

Only at visible and immediately adjacent frequencies are any significant differences in skin reflectivity manifest. People of Northern European ancestry and people of Central African ancestry are equally black in the ultraviolet and in the infrared, where nearly all organic molecules, not just melanin, absorb light. Only in the visible, where many molecules are transparent, is the anomaly of white skin even possible. Over most of the spectrum, all humans are black.*

Sunlight is composed of a mixture of waves with frequencies corresponding to all the colors of the rainbow. There is slightly more yellow light than red or blue, which is partly why the Sun looks yellow. All of these colors fall on, say, the petal of a rose. So why does the rose look red? Because all colors other than red are preferentially absorbed inside the petal. The mixture of light waves strikes the rose. The waves are bounced around helter-skelter below the petal's surface. As with a wave in the bathtub, after every bounce the wave is weaker. But blue and yellow waves are absorbed at each reflection more than red waves. The net result after many interior bounces is that more red light is reflected back than light of any other color, and it is for this reason that we perceive the beauty of a red rose. In blue or violet flowers exactly the same thing happens, except now red and yellow light is preferentially absorbed after multiple interior bounces and blue and violet light is preferentially reflected.

*These are among the reasons that "African-American" (or equivalent hyphenations in other countries) is a much better descriptive than "black" or — the same word in Spanish — "Negro."

There's a particular organic pigment responsible for the absorption of light in such flowers as roses and violets — flowers so strikingly colored that they're named after their hues. It's called anthocyanin. Remarkably, a typical anthocyanin is red when placed in acid, blue in alkali, and violet in water. Thus, roses are red because they contain anthocyanin and are slightly acidic; violets are blue because they contain anthocyanin and are slightly alkaline. (I've been trying to use these facts in doggerel, but with no success.)

Blue pigments are hard to come by in Nature. The rarity of blue rocks or blue sands on Earth and other worlds is an illustration of this fact. Blue pigments have to be fairly complicated; the anthocyanins are composed of about 20 atoms, each heavier than hydrogen, arranged in a particular pattern.

Living things have inventively put color to use — to absorb sunlight and, through photosynthesis, to make food out of mere air and water; to remind mother birds where the gullets of their fledglings are; to interest a mate; to attract a pollinating insect; for camouflage and disguise; and, at least in humans, out of delight in beauty. But all this is possible only because of the physics of stars, the chemistry of air, and the elegant machinery of the evolutionary process, which has brought us into such superb harmony with our physical environment.

And when we're studying other worlds, when we're examining the chemical composition of their atmospheres or surfaces — when we're struggling to understand why the high haze of Saturn's moon Titan is brown and the cantalouped terrain of Neptune's moon Triton pink — we're relying on the properties of light waves not very different from the ripples spreading out in the bathtub. Since all the colors that we see — on Earth and everywhere else — are a matter of which wavelengths of sunlight are best reflected, there is still more than poetic merit to think of the Sun as caressing all within its reach, of sunlight as the gaze of God. But you have a much better shot at understanding what's happening if you think instead of a dripping faucet.

Chapter 5

FOUR COSMIC QUESTIONS

When on high the heaven had not been named,
Firm ground below had not been called by name . . .
No reed hut had been matted, no marsh land had appeared,
When no god whatever had been brought into being,
Uncalled by name, their destinations undetermined —
Then it was that the gods were formed . . .

Enuma Elish,
the Babylonian creation myth
(late third millennium B.C.)*

*"Enuma elish" are the first words of the myth, as if the Book of Genesis were called "In the Beginning" — which is in fact nearly the meaning of the Greek word "genesis."

Every culture has its creation myth — an attempt to understand where the Universe came from, and all within it. Almost always these myths are little more than stories made up by story tellers. In our time, we have a creation myth also. But it is based on hard scientific evidence. It goes something like this . . .

We live in an expanding Universe, vast and ancient beyond ordinary human understanding. The galaxies it contains are rushing away from one another, the remnants of an immense explosion, the Big Bang. Some scientists think the Universe may be one of a vast number — perhaps an infinite number — of other closed-off universes. Some may grow and then collapse, live and die, in an instant. Others may expand forever. Some may be poised delicately and undergo a large number — perhaps an infinite number — of expansions and contractions. Our own Universe is about 15 billion years past its origin, or at least its present incarnation, the Big Bang.

There may be different laws of Nature and different forms of matter in those other universes. In many of them life may be impossible, there being no suns and planets, or even no chemical elements more complicated than hydrogen and helium. Others may have an intricacy, diversity, and richness that dwarfs our own. If those other universes exist, we may never be able to plumb their secrets, much less visit them. But there is plenty to occupy us about our own.

Our Universe is composed of some hundred billion galaxies, one of which is the Milky Way. "*Our* Galaxy," we like to call it, although we certainly do not have possession of it. It is composed of gas and dust and about 400 billion suns. One of them, in an obscure spiral arm, is the Sun, the local star — as far as we can tell, drab, humdrum, ordinary. Accompanying the Sun in its 250 million year journey around the center of the Milky Way is a retinue of small worlds. Some are planets, some are moons, some asteroids, some comets. We humans are one of the 50 billion species that have grown up and evolved on a small planet, third from the Sun, that we call the Earth. We have sent spacecraft to examine seventy of the other worlds in our system, and to enter the atmospheres or land on the surfaces of four of them — the Moon, Venus, Mars, and Jupiter. We have been engaged in a mythic endeavor.

Prophecy is a lost art. Despite our "eager desire to pierce the thick darkness of futurity," in Charles McKay's words, we're often not very good at it. In science the most important discoveries are often the most unexpected — not a mere extrapolation from what we currently know, but something completely different. The reason is that Nature is far more inventive, subtle, and elegant than humans are. So in a way it's foolish to attempt to anticipate what the most significant findings in astronomy might be in the next few decades, the future adumbration of our creation myth. But on the other hand, there are discernible trends in the development of new instrumentation that indicate at least the prospect of goose-bump-raising new discoveries.

Any astronomer's choice of the four most interesting problems will be idiosyncratic and I know many would make choices different from mine. Among other candidate mysteries are what 90 percent of the Universe is made of (we still don't know); identification of the nearest black hole; the bizarre putative result that the distances of galaxies are quantized — that is, galaxies are at certain distances and their multiples but not at intermediary distances; the nature of gamma ray bursters, in which the equivalent of whole solar systems episodically blow up; the apparent paradox that the age of the

Universe may be less than the age of the oldest stars in it (probably resolved by the recent conclusion, using Hubble Space Telescope data, that the Universe is 15 billion years old); the investigation in Earth laboratories of returned cometary samples; the search for interstellar amino acids; and the nature of the earliest galaxies.

Unless there are major cuts in the funding for astronomy and space exploration worldwide — a doleful possibility by no means unthinkable — here are four questions* of enormous promise:

1. **Was There Ever Life on Mars?** The planet Mars is today a bone-dry frozen desert. But all over the planet there are clearly preserved ancient river valleys. There are also signs of ancient lakes and perhaps even oceans. From how cratered the terrain is, we can make a rough estimate of when Mars was warmer and wetter. (The method has been calibrated by cratering on our Moon and radioactive dating from the half-lives of elements in lunar samples returned by *Apollo* astronauts.) The answer is about 4 billion years ago. But 4 billion years ago is just the epoch in which life was arising on Earth. Is it possible that there were two

*A fifth is described in the following chapter.

nearby planets with very similar environments, and life arose on one but not the other? Or did life arise on early Mars, only to be wiped out when the climate mysteriously changed? Or might there be oases or refuges, perhaps subsurface, where some forms of life linger into our own time? Mars thus raises two fundamental enigmas for us — the possible existence of past or present life, and the reason that an Earth-like planet has become locked into a permanent ice age. This latter question may be of practical interest to us, a species that is busily pushing and pulling on its own environment with a very poor understanding of the consequences.

When *Viking* landed on Mars in 1976, it sniffed the atmosphere, finding many of the same gases as in the Earth's atmosphere — carbon dioxide, for example — and a paucity of gases prevalent in the Earth's atmosphere — ozone, for example. What's more, the particular variety of molecule, its isotopic composition, was determined and was in many cases different from the isotopic composition of the comparable molecules on Earth. We had discovered the characteristic signature of the Martian atmosphere.

A curious fact then transpired. Meteorites

— rocks from space — had been found in the Antarctic ice sheet, sitting directly on top of the frozen snows. Some had been discovered by the time of *Viking*, some after; all had fallen to Earth before the *Viking* mission, often tens of thousands of years before. On the clean Antarctic ice shelf, they were not difficult to discern. Most of the meteorites so collected were brought to what in the *Apollo* days had been the Lunar Receiving Laboratory in Houston.

But funding is very meager at NASA these days, and not even a preliminary look at all these meteorites had been performed for years. Some turned out to be from the Moon — a meteorite or comet impacting the Moon, spraying Moon rocks out into space, one or some of which land in Antarctica. One or two of these meteorites come from Venus. And astonishingly, some of them, judging by the Martian atmospheric signature hidden away in their minerals — come from Mars.

In 1995–96, scientists at NASA's Johnson Space Flight Center finally got around to examining one of the meteorites — ALH84001 — that proved to come from Mars. It looked in no way extraordinary, resembling a brownish potato. When the microchemistry was examined, certain species of organic molecules were discovered, chiefly polycyclic aromatic hydrocarbons (PAHs). These are not in themselves all that remarkable. Structurally they resemble the hexagonal patterns on bathroom tiles with a carbon atom at each vertex. PAHs are known in ordinary meteorites, in interstellar grains, and are suspected on Jupiter and Titan. They do not by any means indicate life. But the PAHs were arranged so that there were more of them deeper in the Antarctic meteorite, suggesting that this was not contamination from Earthly rocks (or automobile exhaust), but intrinsic to the meteorite. Still, PAHs in uncontaminated meteorites do not indicate life. Other minerals sometimes associated with life on Earth were also found. But the most provocative result was the discovery of what some scientists are calling nanofossils — tiny spheres attached one to another, like very small bacterial colonies on Earth. But can we be sure that there are no terrestrial or Martian minerals that have a similar form? Is the evidence adequate? For years I've been stressing with regard to UFOs that extraordinary claims require extraordinary evidence. The evidence for life on Mars is not yet extraordinary enough.

But it's a start. It points us to other parts of this particular Martian meteorite. It guides us to other Martian meteorites. It suggests the search for quite different meteorites in the Antarctic ice field. It hints that we search not just for other deeply buried rocks obtained from or on Mars, but for much shallower rocks. It urges upon us a reconsideration of the enigmatic results from the biology experiments on *Viking*, some of which were argued by a few scientists to indicate the presence of life. It suggests sending spacecraft missions to special locales on Mars which may have been the last to surrender their warmth and wetness. It opens up the entire field of Martian exobiology.

And if we are so lucky as to find even a simple microbe on Mars, we have the wonderful circumstance of two nearby planets, each with life on it in the same early epoch. True, maybe life was transported by meteorite impact from one world to another and does not indicate independent origins on each world. We should be able to check that by checking the organic chemistry and morphology of the life-forms uncovered. Maybe life arose on only one of these worlds, but evolved separately on both. We then would have an example of several billion years of independent evolution, a biological bonanza available in no other way.

And if we are most lucky, we will find really independent life-forms. Are they based on nucleic acids for their genetic coding? Are they based on proteins for their enzymatic catalysis? What genetic code do they use? Whatever the answers to these questions, the entire science of biology is the winner. And whatever the outcome, the implication is that life may be much more widespread than most scientists had thought.

In the next decade there are vigorous plans by many nations for robot orbiters, landers, roving vehicles, and subsurface penetrator spacecraft to be sent to Mars to lay the groundwork for answering these questions; and — maybe — in 2005 a robotic mission to return surface and subsurface samples from Mars to Earth.

2. **Is Titan a Laboratory for the Origin of Life?** Titan is the big moon of Saturn, an extraordinary world with an atmosphere ten times denser than the Earth's and made mainly of nitrogen (as here) and methane (CH_4). The two U.S. *Voyager* spacecraft detected a number of simple organic molecules in the atmosphere of Titan — carbon-based compounds that have been implicated in the origin of life on Earth. This moon is surrounded by an

opaque reddish haze layer, which has properties identical to a red-brown solid made in the laboratory when energy is supplied to a simulated Titan atmosphere. When we analyze what this stuff is made of we find many of the essential building blocks of life on Earth. Because Titan is so far from the Sun, any water there should be frozen — and so you might think it is at best an incomplete analog of the Earth at the time of the origin of life. However, occasional impacts by comets are capable of melting the surface, and it looks as if an average place on Titan has been underwater for a millennium, more or less, in its 4.5 billion year history. In the year 2004, a NASA spacecraft called *Cassini* will arrive in the Saturn system; an entry probe built by the European Space Agency called *Huygens* will detach itself and slowly sink through the atmosphere of Titan toward its enigmatic surface. We may then learn how far Titan has gone on the path to life.

3. **Is There Intelligent Life Elsewhere?** Radio waves travel at the speed of light. Nothing goes faster. At the right frequency they pass cleanly through interstellar space and through planetary atmospheres. If the largest radio/radar telescope on Earth were pointing at an equivalent telescope on a

planet of another star, the two telescopes could be separated by thousands of light-years and still hear each other. For these reasons, existing radio telescopes are being used to see if anyone is sending us a message. So far we have found nothing certain, but there have been tantalizing "events" — signals recorded that satisfy all the criteria for extraterrestrial intelligence, except one: You turn the telescope back and point at that patch of sky again, minutes later, months later, years later; and the signal never repeats. We are only at the beginning of the search program. A really thorough search would take a decade or two. If extraterrestrial intelligence is found, then our view of the Universe and ourselves is changed forever. And if after a long and systematic search we find nothing, then we may have calibrated something of the rarity and preciousness of life on Earth. Either way, this is a search well worth doing.

4. **What Is the Origin and Fate of the Universe?** Astonishingly, modern astrophysics is on the verge of determining fundamental insights on the origin, nature, and fate of the entire Universe. The Universe is expanding; all the galaxies are running away from each other in what is called the Hubble flow, one of three main pieces of evi-

dence for an enormous explosion at the time the Universe began — or at least its present incarnation. The gravity of the Earth is strong enough to pull back a stone thrown up into the sky, but not a rocket at escape velocity. And so it is with the Universe: If it contains a great deal of matter, the gravity exercised by all this matter will slow down and stop the expansion. An expanding Universe will be converted into a collapsing Universe. And if there is not enough matter, the expansion will continue forever. The present inventory of matter in the Universe is insufficient to slow the expansion, but there are reasons to think that there may be a great deal of dark matter that does not betray its existence by giving off light for the convenience of astronomers. If the expanding Universe turns out to be only temporary, ultimately to be replaced by a contracting Universe, this would certainly raise the possibility that the Universe goes through an infinite number of expansions and contractions and is infinitely old. An infinitely old Universe has no need to be created. It was always here. If, on the other hand, there is insufficient matter to reverse the expansion, then this would be consistent with a Universe created from nothing. These are deep and difficult questions which every human culture has one way or another tried to grapple with. But only in our time do we have a real prospect of uncovering some of the answers. Not by guesses or stories — but by real, repeatable, verifiable observations.

I think there is a reasonable chance that startling revelations in all four of these areas can be expected in the next decade or two. Again, there are many other questions in modern astronomy that I could have substituted, but the prediction I can make with the highest confidence is that the most amazing discoveries will be ones we are not today wise enough to foresee.

Chapter 6

SO MANY SUNS, SO MANY WORLDS

> What a wonderful and amazing scheme have we here of the magnificent vastness of the Universe! So many Suns, so many Earths . . . !
>
> CHRISTIAN HUYGENS,
> *New Conjectures Concerning the Planetary Worlds, Their Inhabitants and Productions*
> (ca. 1670)

In December 1995, an entry probe, detached from the *Galileo* Jupiter orbiter, entered the turbulent, roiling atmosphere of Jupiter and sank to a fiery death. Along the way it radioed back information on what it found. Four previous spacecraft had examined Jupiter as they raced by. The planet has also been studied by ground-based and space telescopes. Unlike the Earth, which is made mainly of rock and metal, Jupiter is made mostly of hydrogen and helium. It is so big that a thousand Earths could fit inside. At depth, its atmospheric pressure gets so large that electrons are squeezed off atoms and the hydrogen becomes a hot metal. This state of affairs is thought to be the reason that twice as much energy comes pouring out of Jupiter than Jupiter gets from the Sun. The winds that buffeted the *Galileo* probe at its deepest entry point probably arise not from sunlight but from the energy originating in the deep interior. At the very core of Jupiter there seems to be a rocky and iron world many times the mass of the Earth, surmounted by the immense ocean of hydrogen and helium. Visiting the metallic hydrogen — much less the rocky core — is beyond human abilities for at least centuries or millennia to come.

The pressures are so great in the interior of Jupiter that it is hard to imagine life there — even life very different from our own. A few scientists, myself among them, have tried, just for fun, to imagine an ecology that might evolve in the atmosphere of a Jupiter-like planet, somewhat like the microbes and fish in the Earth's oceans. The origin of life might be difficult in such an environment, but we now know that asteroidal and cometary impacts transfer surface material from world to world, and it is even possible that impacts in the early history of the Earth transferred primitive life from our planet to Jupiter. This, though, is the merest speculation.

Jupiter is about 5 astronomical units from the Sun. An astronomical unit (abbreviated AU) is the distance of the Earth from the Sun, about 93 million miles, or 150 million kilometers. If not for the interior heat and the greenhouse effect in Jupiter's immense atmosphere, the temperatures there would be about 160 degrees below zero Celsius. That's roughly the temperature on the surface of Jupiter's moons — much too cold for life.

Jupiter and most of the other planets in our Solar System orbit the Sun in the same plane, as if they were confined to separate grooves on a phonograph record or a compact disc. Why should this be? Why shouldn't the orbital planes be tilted at all angles? Isaac Newton, the mathematical genius who first understood how gravity makes the planets move, was puzzled by the absence of much tilt in the orbital planes of the planets, and deduced that, at the beginning of the Solar System, God must have started the planets out all orbiting in the same plane.

But the mathematician Pierre Simon, the

Marquis de Laplace, and later the celebrated philosopher Immanuel Kant, discovered how it could have happened without recourse to divine intervention. Ironically, they relied on the very laws of physics that Newton had discovered. A brief rendition of the Kant-Laplace hypothesis goes as follows: Imagine an irregular, slowly rotating cloud of gas and dust sitting between the stars. There are many such clouds. If its density is sufficiently high, the gravitational attraction of the various parts of the cloud for each other will overwhelm the internal random motion, and the cloud will start contracting. As it does so, it will spin faster, like a pirouetting ice skater bringing in her arms. The spin won't retard the collapse of the cloud along the axis of rotation, but it will slow the contraction down in the plane of rotation. The initially irregular cloud converts itself into a flat disk. So planets that accrete or condense out of this disk will all be orbiting pretty much in the same plane. The laws of physics suffice, without supernatural intervention.

But predicting that such a disklike cloud existed before the planets formed is one thing; confirming the prediction by actually seeing such disks around other stars is quite another. When other spiral galaxies like the

Milky Way were discovered, Kant thought that *these* were the predicted preplanetary disks, and that the "nebular hypothesis" of the origin of planets had been confirmed. (*Nebula* comes from the Greek word for cloud.) But these spiral forms proved to be distant star-studded galaxies and not nearby birthing grounds of stars and planets. Circumstellar disks proved hard to find.

It was not until more than a century later, using equipment including orbiting observatories, that the nebular hypothesis was confirmed. When we look at young Sun-like stars, like our Sun of four or five billion years ago, we find that more than half of them are surrounded by flat disks of dust and gas. In many cases the parts close to the star seem to be empty of dust and gas, as if planets had already formed there, gobbling up the interplanetary matter. It is not definitive evidence, but it strongly suggests that stars like our own frequently, if not invariably, are accompanied by planets. Such discoveries expand the likely number of planets in the Milky Way Galaxy at least into the billions.

But what about actually detecting other planets? Granted, the stars are very far away — the nearest almost a million AU distant — and in visible light they shine only in reflection. But our technology is improving by leaps and bounds. Shouldn't we be able to detect at least large cousins of Jupiter around nearby stars, perhaps in infrared if not visible light?

In the last few years we have entered into a new era in human history, where we are able to detect the planets of other stars. The first planetary system reliably discovered accompanies a most unlikely star: B 1257 + 12 is a rapidly rotating neutron star, the remnant of a star once more massive than the Sun that blew itself up in a colossal supernova explosion. The magnetic field of this neutron star captures electrons and constrains them to move in such paths that, like a lighthouse, they shine a beam of radio light across interstellar space. By chance, the beam intercepts the Earth — once every 0.0062185319388187 seconds. This is why B 1257 + 12 is called a pulsar. The constancy of its period of rotation is astonishing. Because of the high precision of the measurements, Alex Wolszczan, now at Penn State University, was able to find "glitches" — irregularities in the last few decimal places. What causes them? Starquakes or other phenomena on the neutron star itself? Over the years, they have varied in precisely the way expected were there planets going around B 1257 + 12, tugging slightly, first this way and then that. The quantitative agreement is so exact that the conclusion is compelling: Wolszczan has discovered the first planets known beyond the Sun. What's more, they're not big Jupiter-sized planets. Two of them are probably only a little more massive than the Earth, and orbit their star at distances not too different from the Earth's distance from the Sun, 1 AU. Might we expect life on these planets? Unfortunately, there is a gale of charged particles hurtling out of the neutron star, which will raise the temperature of its Earth-like planets far above the boiling point of water. At 1,300 light-years away we will not be traveling to this system soon. It is a current mystery whether these planets survived the supernova explosion that made the pulsar, or were formed from the debris of the supernova explosion.

Shortly after Wolszczan's epochal discovery, several more objects of planetary mass were discovered (mainly by Geoff Marcy and Paul Butler of San Francisco State University) going around other stars — in this case, ordinary Sun-like stars. The technique used was different and much more difficult to apply. These planets were found by conventional optical telescopes monitoring the periodic changes in the spectra of nearby stars. Sometimes a star may be moving for a while toward us and then away from us, as determined by the changes in wavelength of its spectral lines, the Doppler Effect — akin to the changes in frequency of a car's horn as it drives toward or away from us. Some invisible body is tugging at the star. Again, an unseen world is discovered by a quantitative agreement — between the observed slight periodic motions of the star and what you would expect if the star had a nearby planet.

The planets responsible go around the stars 51 Pegasi, 70 Virginis, and 47 Ursae Majoris, respectively in the constellations Pegasus, Virgo, and Ursa Major, the Big Dipper. In 1996, such planets were also found orbiting the star 55 Cancri in the constellation Cancer, the Crab: Tau Bootis and Upsilon Andromedae. Both 47 Ursae Majoris and 70 Virginis can be seen with the naked eye in the spring evening sky. They are very near as stars go. The masses of these planets seem to range from a little less than Jupiter to several times more than Jupiter. What is most surprising about them is how close to their stars they are, from 0.05 AU for 51 Pegasi, to a little more than 2 AU for 47 Ursae Majoris. These systems may also contain smaller Earth-like planets, not yet discovered, but their layout is not like ours.

In our Solar System, we have the small Earth-like planets on the inside and the large Jupiter-like planets on the outside. For these four stars, the Jupiter-mass planets seem to be on the inside. How that could be, no one now understands. We do not even know that these are truly Jupiter-like planets, with immense atmospheres of hydrogen and helium, metallic hydrogen down deep and an Earth-like core still deeper. But we do know that the atmospheres of Jupiter-like planets at such close distances to their stars will not evaporate away. It seems implausible that they formed in the periphery of their solar systems and somehow wandered much closer to their stars. But maybe some early massive planets could have been slowed by the nebular gas and spiraled in. Most experts hold that a Jupiter could not be formed so close to the star.

Why not? Our standard understanding of the origin of Jupiter is something like this: In the outer parts of the nebular disk, where the temperatures were very low, worldlets of ice and rock condensed out, something like the comets and icy moons in the outer parts of our Solar System. These frigid worldlets collided at low speeds, stuck together, and gradually became large enough to gravitationally attract the prevalent hydrogen and helium gases from the nebula, forming a Jupiter from the inside out. In contrast, nearer to the star, it is thought, the nebular temperatures were too high for ice to condense in the first place, and the whole process is short-circuited. But I wonder if some nebular disks were below the freezing point of water even very close to the local star.

In any case, with Earth-mass planets around a pulsar and four new Jupiter-mass planets about Sun-like stars, it follows that our kind of solar system may hardly be typical. This is key if we have any hopes of constructing a general theory of the origin of planetary systems: It now must encompass a diversity of planetary systems.

Still more recently, a technique called astrometry has been used to detect two and possibly three Earth-like planets around a star very near to our Sun, Lalande 21185. Here the precise motion of the star is monitored over many years, and the recoil due to any planets in orbit about it is carefully watched. Departures from circular or elliptical orbits by Lalande 21185 permit us to detect the presence of planets. So here we have a familiar, or at least a somewhat familiar, planetary system to our own. There seem to be at least two and maybe more categories of planetary systems in adjacent interstellar space.

As for life on these Jupiter-like worlds, it is no more likely than on our own Jupiter. But what is probable is that these other Jupiters have moons, like the 16 that circle our Jupiter. Because these moons, like the giant worlds they orbit, are close to the local star, their temperatures, especially for 70 Virginis, might be clement for life. At 35 to 40 light-years away, these worlds are close enough for us to begin to dream of one day sending very fast spacecraft to visit them, the data to be received by our descendants.

Meanwhile, a range of other techniques are coming along. Besides pulsar timing glitches and Doppler measurements of the radial velocities of stars, interferometers on the ground or, better, in space; ground-based telescopes that cancel out the turbulence of the Earth's atmosphere; ground-based observations using the gravitational lens effect of distant massive objects; and very accurate space-borne measurements of the dimming of a star when one of its planets passes in front of it all seem ready in the next few years to yield significant results. We are now on the verge of trolling through thousands of nearby stars, searching for their companions. To me it seems likely that in the coming decades we will have information on at least hundreds of other planetary systems close to us in the vast Milky Way Galaxy — and perhaps even a few small blue worlds graced with water oceans, oxygen atmospheres, and the telltale signs of wondrous life.

Part II

WHAT ARE CONSERVATIVES CONSERVING?

Chapter 7

THE WORLD THAT CAME IN THE MAIL

The world? Moonlit
Drops shaken
From the crane's bill.

DOGEN (1200–1253),
"Wake on Impermanence," from
Lucien Stryk and Takashi Ikemoto,
Zen Poems of Japan: The Crane's Bill
(New York: Grove Press, 1973)

The world arrived in the mail. It was marked "Fragile." A sticker depicting a cracked goblet was on the wrapping. I opened it carefully, dreading the tinkle of broken crystal, or the discovery of a shard of glass. But it was intact. With both hands, I lifted it out and held it up to the sunlight. It was a transparent sphere, about half filled with water. The number 4210 was inconspicuously taped to it. World number 4210: There must be many such worlds. Cautiously, I placed it on the accompanying Lucite stand and peered in.

I could see life in there — a network of branches, some encrusted with green filamentous algae, and six or eight small animals, mostly pink, cavorting, so it seemed, among the branches. In addition, there were hundreds of other kinds of beings, as plentiful in these waters as fish in the oceans of Earth; but they were all microbes, much too small for me to see with the naked eye. Clearly, the pink animals were shrimp of some suitably unpretentious variety. They caught your attention immediately because they were so *busy*. A few had alighted on branches and were walking on 10 legs and waving lots of other appendages. One of them was devoting all its attention, and a considerable number of limbs, to dining on a filament of green. Among the branches, draped with algae much as trees in Georgia and north Florida are covered with Spanish moss, other shrimp could be seen moving as if they had urgent appointments elsewhere. Sometimes they would change their colors as they swam from environment to environment. One was pale, almost transparent; another orange, with an embarrassed blush of red.

In some ways, of course, they were different from us. They had their skeletons on the outside, they could breathe water, and a kind of anus was located disconcertingly near their mouths. (They were fastidious about appearance and cleanliness, though, possessing a pair of specialized claws with brushlike bristles. Occasionally, one would give itself a good scrub.)

But in other ways they were like us. It was hard to miss. They had brains, hearts, blood, and eyes. That flurry of swimming appendages propelling them through the water betrayed what seemed to be an unmistakable hint of purpose. When they arrived at their destination, they addressed the algal filaments with the precision, delicacy, and industriousness of a dedicated gourmet. Two of them, more venturesome than the rest, prowled this world's ocean, swimming high above the algae, languidly surveying their domain.

After a while you get so you can distinguish individuals. A shrimp will molt, shedding its old skeleton to make room for a new one. Afterward, you can see the thing — transparent, shroudlike, hanging rigidly from a branch, its former occupant going about his business in a sleek new carapace. Here's one missing a leg. Had there been some furious claw-to-claw combat, perhaps over the affections of a devastating nubile beauty?

From certain angles, the top of the water is a mirror, and a shrimp sees its own reflection. Can it recognize itself? More probably, it just sees the reflection as one more shrimp. At other angles, the thickness of the curved glass magnifies them, and then I can make out what they really look like. I notice, for example, that they have mustaches. Two of them race to the top of the water and, unable to break through the surface tension, bounce off the meniscus. Then, upright — a little startled, I imagine — they gently sink to the bottom. Their arms are crossed casually, it almost seems, as if the exploit were routine, nothing to write home about. They're cool.

If I can clearly see a shrimp through the curved crystal, I figure, it must be able to see me, or at least my eye — some great looming black disk, with a corona of brown and green. Indeed, sometimes as I watch one busily fingering the algae, it seems to stiffen and look back at me. We have made eye contact. I wonder what it thinks it sees.

After a day or two of preoccupation with work, I wake up, take a glance at the crystal world. . . . They all seem to be gone. I reproach myself. I'm not required to feed them or give them vitamins or change their water or take them to the vet. All I have to do is make sure that they're not in too much light or too long in the dark and that they're always at temperatures between 40° and 85° Fahrenheit. (Above that, I guess, they make a bisque and not an ecosystem.) Through inattention, have I killed them? But then I see one poking an antenna out from behind a branch, and I realize they're still in good health. They're only shrimp, but after a while you find yourself worrying about them, rooting for them.

If you're in charge of a little world like this, and you conscientiously concern yourself about its temperature and light levels, then — whatever you may have had in mind at the beginning — eventually you care about who's *in* there. If they're sick or dying, though, you can't do much to save them. In certain ways, you're much more powerful than they, but they do things — like breathing water — that you can't. You're limited, painfully limited. You even wonder if it's cruel to put them in this crystal prison. But you reassure yourself that at least here they are safe from baleen whales and oil slicks and cocktail sauce.

The ghostly molting shrouds and the rare dead body of an expired shrimp do not linger long. They are eaten, partly by the other shrimp, partly by invisible microorganisms that teem through this world's ocean. And so you are reminded that these creatures don't work by themselves. They *need* one another. They look after one another — in a way that I'm unable to do for them. The shrimp take oxygen from the water and exhale carbon dioxide. The algae (a plural word; singular: alga) take carbon dioxide from the water and exhale oxygen. They breathe each other's waste gases. Their solid wastes cycle also, among plants and animals and microorganisms. In this small Eden, the inhabitants have an extremely intimate relationship.

The shrimps' existence is much more tenuous and precarious than that of the other beings. The algae can live without the shrimp far longer than the shrimp can live without the algae. The shrimp eat the algae, but the algae mainly eat light. Eventually — to this day I don't know why — the shrimp started dying, one by one. The time came when only one was left, morosely — it seemed — nibbling a sprig of algae until it too died. A little to my surprise, I found I was mourning them all. I suppose it was in part because I had gotten to know them a little. But in part, I knew, it was because I feared a parallelism between their world and ours.

Unlike an aquarium, this little world is a closed ecological system. Light gets in, but nothing else — no food, no water, no nutrients. Everything must be recycled. Just like the Earth. In our larger world, we also — plants and animals and microorganisms — live off each other, breathe and eat each other's wastes, depend on one another. Life on our world, too, is powered by light. Light from the Sun, which passes through the clear air, is harvested by plants and powers them to combine carbon dioxide and water into carbohydrates and other foodstuffs, which in turn provide the staple diet of the animals.

Our big world is very like this little one, and we are very like the shrimp. But there is at least one major difference: Unlike the shrimp, we are able to change our environment. We can do to ourselves what a careless owner of such a crystal sphere can do to the shrimp. If we are not careful, we can heat our planet through the atmospheric greenhouse effect or cool and darken it in the aftermath of a nuclear war or the massive burning of an oil field (or by ignoring the danger of an asteroid or cometary impact). With acid rain, ozone depletion, chemical pollution, radioactivity, the razing of tropical forests, and a dozen other assaults on the environment, we are pushing and pulling our little world in poorly understood directions. Our purportedly advanced civilization may be changing the delicate ecological balance that has tortuously evolved over the 4-billion-year period of life on Earth.

Crustacea, such as shrimp, are much older than people or primates or even mammals. Algae go back 3 billion years, long before animals, most of the way back to the origin of life on Earth. They've all been working together — plants, animals, microbes — for a very long time. The arrangement of organisms in my crystal sphere is ancient, vastly older than any cultural institution we know. The inclination to cooperate has been painfully extracted through the evolutionary process. Those organisms that did not cooperate, that did not work with one another, died. Cooperation is encoded in the survivors' genes. It's their *nature* to cooperate. It's a key to their survival.

But we humans are newcomers, arising only a few million years ago. Our present technical civilization is just a few hundred years old. We have not had much recent experience in voluntary interspecies (or even intraspecies) cooperation. We are very devoted to the short-term and hardly ever think about the long-term. There is no guarantee that we will be wise enough to understand our planetwide closed ecological system, or to modify our behavior in accord with that understanding.

Our planet is indivisible. In North America, we breathe oxygen generated in the Brazilian rain forest. Acid rain from polluting industries in the American Midwest destroys Canadian forests. Radioactivity from a Ukrainian nuclear accident compromises the economy and culture of Lapland. The burning of coal in China warms Argentina. Chlorofluorocarbons released from an air conditioner in Newfoundland help cause skin cancer in New Zealand. Diseases rapidly spread to the farthest reaches of the planet and require a global medical effort to be eradicated. And, of course, nuclear war and asteroid impact imperil everyone. Like it or not, we humans are bound up with our fellows, and with the other plants and animals all over the world. Our lives are intertwined.

If we are not graced with an instinctive knowledge of how to make our technologized world a safe and balanced ecosystem, we must *figure out* how to do it. We need more scientific research and more technological restraint. It is probably too much to hope that some great Ecosystem Keeper in the sky will reach down and put right our environmental abuses. It is up to us.

It should not be impossibly difficult. Birds — whose intelligence we tend to malign — know not to foul the nest. Shrimps with brains the size of lint particles know it. Algae know it. One-celled microorganisms know it. It is time for us to know it too.

Chapter 8

THE ENVIRONMENT: WHERE DOES PRUDENCE LIE?

> This new world may be safer, being told
> The dangers of diseases of the old
>
> JOHN DONNE,
> "An Anatomie of the World — The First Anniversary" (1611)

There's a certain moment at twilight when the aircraft contrails are pink. And if the sky is clear, their contrast with the surrounding blue is unexpectedly lovely. The Sun has already set and there's a roseate glow at the horizon, a reminder of where the Sun is hiding. But the jet aircraft are so high up that *they* can still see the Sun — quite red, just before setting. The water blown out from their engines instantly condenses. At the frigid temperatures of high altitude, each engine trails a small, linear cloud, illuminated by the red rays of the setting Sun.

Sometimes there are several contrails from different aircraft, and they cross, making a kind of sky writing. When the winds are high, the contrails quickly spread laterally, and instead of an elegant line tracing its way across the sky, there's a long, irregular, diffuse, vaguely linear tracery that dissipates as you watch. If you catch the trail as it's being generated, you can often make out the tiny object from which it emanates. For many people, no wings or engines are visible; it's just a moving spot separated a little from the contrail, somehow its source.

As it gets darker you can often see that the spot is self-luminous. There's a bright white light there. Sometimes there's also a flashing red or green light, or both.

Occasionally I imagine myself a hunter-gatherer — or even my grandparents when they were children — looking up at the sky and seeing these bewildering and awesome wonders from the future. For all the days of humans on Earth, it is only in the twentieth century that we have become a presence in the sky. While the air traffic in upstate New York, where I live, is doubtless heavier than in many places on Earth, there is hardly anywhere on the planet where you can't, at least occasionally, look up and see our machines writing out their mysterious messages on the very sky that we had so long thought of as the exclusive provenance of the gods. Our technology has reached astonishing proportions for which, in our heart of hearts, we are inadequately prepared, mentally or emotionally.

A little later, when the stars begin to come out, I can make out among them an occasional moving bright light, sometimes quite bright. Its glow may be steady, or it may be blinking at me, often two lights in tandem. No longer are there cometlike tails trailing behind them. There are moments when 10 or 20 percent of the "stars" I can see are nearby artifacts of humanity, which can for a moment be confused with immensely distant, blazing suns. More rarely, well after sunset, I can see a point of light, usually quite dim, very slowly and subtly moving. I have to be sure that first it passes this star, and then that — because the eye has a penchant for thinking that any isolated point of light surrounded only by blackness is moving. These are not aircraft. They are spacecraft. We have made machines that circle the Earth once every hour and a half. If they're especially big or reflective, we can see them with the naked eye. They are far above the atmosphere, in the blackness of nearby space. They are so high up that they can see the Sun even when it is nearly pitch-dark down here. Unlike the airplanes, they have no lights of their own. Like the Moon and the planets, they shine merely by reflected sunlight.

The sky begins not too far above our heads. It encompasses both the Earth's thin atmosphere and all the vastness of the Cosmos beyond. We have built machines that fly in these realms. We are so accustomed to this, so acclimatized, that we often fail to recognize what a mythic accomplishment it is. More than any other feature of our technical civilization, these now prosaic flights are emblematic of what powers are now ours.

But with great powers come great responsibilities.

Our technology has become so powerful that — not only consciously, but also inadvertently — we are becoming a danger to ourselves. Science and technology have saved billions of lives, improved the well-being of many more, bound up the planet in a slowly anastomosing unity — and at the same time changed the world so much that many people no longer feel at home in it. We've created a range of new evils: hard to see, hard to understand, problems that cannot readily be cured — certainly not without challenging those already in power.

Here, if anywhere, public understanding of science is essential. Many scientists claim that there are real dangers consequent to our continuing to do things the way we've done, that our industrial civilization is a booby trap. But if we were to take such dire warnings seriously, it would be expensive. The affected industries would lose profits. Our own anxiety would increase. There are natural enough reasons to try to reject the warnings. Maybe the large number of scientists who caution about impending catastrophes are worrywarts. Maybe they get a perverse pleasure out of scaring the rest of us. Maybe it's just a way to pry research money out of the government. After all, there are other scientists who say there's nothing to worry about, that the contentions are unproved, that the environment will heal itself. Naturally we long to believe them; who wouldn't? If they're right, it relieves our burden immensely. So let's not jump into things. Let's be cautious. Let's go slow. Let's be really sure.

On the other hand, maybe those who are reassuring about the environment are Pollyannas, or are afraid to affront those in power, or are supported by those who profit by despoiling the environment. So let's hurry up. Let's fix things before they become unfixable.

How do we decide?

There are arguments and counterarguments concerning abstractions, invisibilities, unfamiliar concepts and terms. Sometimes even words like "fraud" or "hoax" are uttered about the dire scenarios. How good is the science here? How can the average person be informed on what the issues are? Can't we maintain a dispassionate but open neutrality and let the contending parties fight it out, or wait until the evidence is absolutely unambiguous? After all, extraordinary claims require extraordinary evidence. In short, why should those who, like myself, teach skepticism and caution about *some* extraordinary claims argue that other extraordinary claims must be taken seriously and considered urgent?

Every generation thinks its problems are unique and potentially fatal. And yet every generation has survived to the next. Chicken Little, it is suggested, is alive and well.

Whatever merit this argument may once have had — and certainly it provides a useful counterbalance to hysteria — its cogency is much diminished today. We sometimes hear about the "ocean" of air surrounding the Earth. But the thickness of most of the atmosphere — including all of it involved in the greenhouse effect — is only 0.1 percent of the diameter of the Earth. Even if we include the high stratosphere, the atmosphere isn't as much as 1 percent of the Earth's diameter. "Ocean" sounds massive, imperturbable. Compared with the size of the Earth, though, the thickness of the air is something like the thickness of the coat of shellac on a large schoolroom globe compared with the globe itself. If the protective ozone layer were brought down from the stratosphere to the surface of the Earth, its thickness compared with the diameter of the Earth would be one part in four billion. It would be utterly invisible. Many astronauts have reported seeing that delicate, thin, blue aura at the horizon of the daylit hemisphere — that represents the thickness of the entire atmosphere — and immediately, unbidden, contemplating its fragility and vulnerability. They worry about it. They have reason to worry.

Today we face an absolutely new circumstance, unprecedented in all of human history. When we started out, hundreds of thousands of years ago, say, with an average population density of a hundredth of a person per square kilometer or less, the triumphs of our technology were hand axes and fire; we were unable to make major changes in the global environment. The idea would never have occurred to us. We were too few and our powers too feeble. But as time went on, as technology improved, our numbers increased exponentially, and now here we are with an average of some ten people per square kilometer, our numbers concentrated in cities, and an awesome technological armory at hand — the powers of which we understand and control only incompletely.

Because our lives depend on minuscule amounts of such gases as ozone, major environmental disruption can be brought about — even on a planetary scale — by the engines of industry. The inhibitions placed on the irresponsible use of technology are weak, often halfhearted, and almost always, worldwide, subordinated to short-term national or corporate interest. We are now able, intentionally or inadvertently, to alter the global environment. Just how far along we are in working the various prophesied planetary catastrophes is still a matter of scholarly debate. But that we are able to do so is now beyond question.

Maybe the products of science are simply too powerful, too dangerous for us. Maybe we're not grown-up enough to be given them. Would it be wise to give a handgun as a present to an infant in the crib? What about a toddler, or a preadolescent child, or a teenager? Or perhaps, as some have argued, automatic weapons should be given to no one in civilian life, because all of us have experienced at one time or another blinding if childish passions. If only the weapon were not around, it so often seems, the tragedy would not have happened. (Of course there are reasons people give for having handguns, and there may be circumstances in which those reasons are valid. Likewise for the dangerous products of science.) Now one further complication: Imagine that when you pull the trigger on a handgun, it takes decades before either the victim or the assailant recognizes that someone's been hit. Then it's even more difficult to grasp the dangers of having weapons around. The analogy is imperfect, but something like this applies to the global environmental consequences of modern industrial technology.

There is, it seems to me, good cause to question, to speak out, to devise new institutions and new ways of thinking. Yes, civility is a virtue and can reach an opponent deaf to the most fervent philosophical entreaties. Yes, it is absurd to try to convert everyone to a new way of thinking. Yes, we might be wrong and our opponent right. (It has been known to happen.) And yes, it is rare that one disputant in an argument convinces another. (Thomas Jefferson said he had never seen it happen, but that seems too harsh. It happens in science all the time.) But these are not adequate reasons to shy from public debate.

Through better medical practice, pharmaceuticals, agriculture, contraception, advances in transportation and communications, devastating new weapons of war, inadvertent side effects of industry, and disquieting challenges to long held worldviews, science and technology have dramatically changed our lives. Many of us are huffing and puffing to keep up, sometimes only slowly grasping the implications of the new developments. In the ancient human tradition, young people grasp change more quickly than the rest of us — not just in running personal computers and programming videocassette recorders, but also in accommodating to new visions of our world and ourselves. The current pace of change is much quicker than a human lifetime, so fast as to work to rend the generations asunder. This middle section of the book is about understanding and accommodating to the environmental upheavals — both for good and for ill — brought on by science and technology.

I will concentrate on the thinning ozone layer and on global warming — as representative of the dilemmas we face. But there are many other worrisome environmental consequences of human technology and expansiveness: rendering vast numbers of species extinct, when desperately needed medicines for cancer, heart disease, and other deadly diseases come from rare or endangered species; acid rain; nuclear, biological, and chemical weapons; and toxic chemicals (and radioactive poisons) — often located in the neighborhoods of the poorest and least powerful among us. An unexpected new finding, disputed by other scientists, is a precipitous recent decline in America, Western Europe, and elsewhere in sperm counts — possibly from chemicals and plastics that mimic the female sex hormones. (The decline is so steep, some say, that, if it continues, men in the West could in consequence start becoming sterile by the middle twenty-first century.)

The Earth is an anomaly. In all the Solar System, it is, so far as we know, the only inhabited planet. We humans are one amongst millions of separate species who live in a world burgeoning, overflowing with life. And yet, most species that ever were are no more. After flourishing for 180 million years, the dinosaurs were extinguished. Every last one. There are none left. No species is guaranteed its tenure on this planet. And we've been here for only about a million years, we, the first species that has devised means for its self-destruction. We are rare and precious because we are alive, because we can think as well as we can. We are privileged to influence and perhaps control our future. I believe we have an obligation to fight for life on Earth — not just for ourselves, but for all those, humans and others, who came before us, and to whom we are beholden, and for all those who, if we are wise enough, will come after. There is no cause more urgent, no dedication more fitting than to protect the future of our species. Nearly all our problems are made by humans and can be solved by humans. No social convention, no political system, no economic hypothesis, no religious dogma is more important.

Everyone experiences at least a dull background level of assorted anxieties. They almost never go away entirely. Most of them are of course about our everyday lives.

There is a clear survival value to this buzz of whispered reminders, wincing recollections of past faux pas, mental testings of possible responses to imminent problems. For too many of us the anxiety is about finding enough for our children to eat. Anxiety is one of those evolutionary compromises — optimized so there will be a next generation, but painful to this generation. The trick, if you can pull it off, is to pick the right anxieties. Somewhere between cheerful dolts and nervous worrywarts there's a state of mind we ought to embrace.

Except for millenarians of the various denominational persuasions and the tabloid press, the only group of people that seems routinely to worry about new claims of disasters — catastrophes unglimpsed in the entire written history of our species — are the scientists. They get to understanding how the world is, and it occurs to them that it might be very different. A little push here, a little tug there, and big changes could happen. Because we humans are generally well adapted to our circumstances — ranging from the global climate to the political climate — any change is likely to be disturbing, painful, and expensive. So naturally we tend to require of the scientists that they be pretty sure of what they're telling us before we run off and protect ourselves against an imaginary danger. Some of the alleged dangers seem so serious, though, that the thought arises unbidden that it may be prudent to take seriously even a small chance of a very grave peril.

The anxieties of everyday life work in a similar way. We buy insurance and caution the children about talking to strangers. For all the anxieties, sometimes we miss the dangers altogether: "Everything I worried about never happened. All the bad things came out of nowhere," one acquaintance told my wife, Annie, and me.

The worse the catastrophe is, the harder it is to keep our balance. We want so badly either to ignore it utterly or to devote all our resources to circumventing it. It's hard soberly to contemplate our circumstances and put the associated anxiety aside for a moment. Too much seems at stake. In the following pages I try to describe some of the current actions of our species that seem disturbing — in how we care for the planet, and how we arrange our politics. I try to show both sides but — I freely admit — I have a point of view deriving from my assessment of the weight of the evidence. Where humans make problems, humans can make solutions, and I've tried to indicate how some of our problems might be solved. You might think a different set of problems should have higher priority, or that there are a different set of solutions. But I hope you'll find in reading this section of the book that you're provoked into contemplating the future a little more. I don't wish unnecessarily to add to our burden of anxieties — almost all of us have a sufficient number — but there are some issues that not enough of us, it seems to me, are thinking through. This sort of contemplating the future consequences of present actions has a proud lineage among us primates, and is one of the secrets of what is still, by and large, the stunningly successful story of humans on Earth.

Chapter 9

CROESUS AND CASSANDRA

It takes courage to be afraid.

MONTAIGNE,
Essays, III, 6 (1588)

Apollo, an Olympian, was god of the Sun. He was also in charge of other matters, one of which was prophecy. That was one of his specialties. Now the Olympian gods could all see into the future a little, but Apollo was the only one who systematically offered this gift to humans. He established oracles, the most famous of which was at Delphi, where he sanctified the priestess. She was called the Pythia, after the python that was one of her incarnations. Kings and aristocrats — and occasionally ordinary people — would come to Delphi and beg to know what was to be.

Among the supplicants was Croesus, King of Lydia. We remember him in the phrase "rich as Croesus," which is still nearly current. Perhaps he has come to be synonymous with wealth because it was in his time and kingdom that coins were invented — minted by Croesus in the seventh century B.C. (Lydia was in Anatolia, contemporary Turkey.) Clay money was a much earlier Sumerian invention. His ambition could not be contained within the boundaries of his small nation. And so, according to Herodotus's *History*, he got it into his head that it would be a good idea to invade and subdue Persia, then the superpower of Western Asia. Cyrus had united the Persians and the Medes and forged a mighty Persian Empire. Naturally, Croesus had some degree of trepidation.

In order to judge the wisdom of invasion, he dispatched emissaries to consult the Delphic Oracle. You can imagine them laden with opulent gifts — which, incidentally, were still on display in Delphi a century later, in Herodotus's time. The question the emissaries put on Croesus's behalf was, "What will happen if Croesus makes war on Persia?"

Without hesitation, the Pythia answered, "He will destroy a mighty empire."

"The gods are with us," thought Croesus, or words to that effect. "Time to invade!"

Licking his chops and counting his satrapies, he gathered his mercenary armies. Croesus invaded Persia — and was humiliatingly defeated. Not only was Lydian power destroyed, but he himself became, for the rest of his life, a pathetic functionary in the Persian court, offering little pieces of advice to often indifferent officials — a hanger-on ex-king. It's a little bit like the Emperor Hirohito living out his days as a consultant on the Beltway in Washington, D.C.

Well, the injustice of it really got to him. After all, he had played by the rules. He had asked for advice from the Pythia, he had paid handsomely, and she had done him wrong. So he sent another emissary to the Oracle (with much more modest gifts this time, appropriate to his diminished circumstances) and asked, "How could you do this to me?" Here, from Herodotus's *History*, is the answer:

> The prophecy given by Apollo ran that if Croesus made war upon Persia, he would destroy a mighty empire. Now in the face of that, if he had been well-advised, he should have sent and inquired again, whether it was his own empire or that of Cyrus that was spoken

of. But Croesus did not understand what was said, nor did he make question again. And so he has no one to blame but himself.

If the Delphic Oracle were only a scam to fleece gullible monarchs, then of course it would have needed excuses to explain away the inevitable mistakes. Disguised ambiguities were its stock in trade. Nevertheless, the lesson of the Pythia is germane: Even of oracles we must ask questions, intelligent questions — even when they seem to tell us exactly what we wish to hear. The policymakers must not blindly accept; they must understand. And they must not let their own ambitions stand in the way of understanding. The conversion of prophecy into policy must be made with care.

This advice is fully applicable to the modern oracles, the scientists and think tanks and universities, the industry-funded institutes, and the advisory committees of the National Academy of Sciences. The policymakers send, sometimes reluctantly, to ask of the oracle, and the answer comes back. These days the oracles often volunteer their prophecies even when no one asks. Their utterances are usually much more detailed than the questions — involving methyl bromide, say, or the circumpolar vortex, hydrochlorofluorocarbons or the West Antarctic Ice Sheet. Estimates are sometimes phrased in terms of numerical probabilities. It seems almost impossible for the honest politician to elicit a simple yes or no. The policymakers must decide what, if anything, to do in response. The first thing to do is to understand. And because of the nature of the modern oracles and their prophecies, policymakers need — more than ever before — to understand science and technology. (In response to this need, the Republican Congress has foolishly abolished its own Office of Technology Assessment. And there are almost no scientists who are members of the U.S. Congress. Much the same is true of other countries.)

But there's another story about Apollo and oracles, at least equally famous, at least equally relevant. This is the story of Cassandra, Princess of Troy. (It begins just before the Mycenaean Greeks invade Troy to start the Trojan War.) She was the smartest and the most beautiful of the daughters of King Priam. Apollo, constantly on the prowl for attractive humans (as were virtually all the Greek gods and goddesses), fell in love with her. Oddly — this almost never happens in Greek myth — she resisted his advances. So he tried to bribe her. But what could he give her? She was already a princess. She was rich and beautiful. She was happy. Still, Apollo had a thing or two to offer. He promised her the gift of prophecy. The offer was irresistible. She agreed. *Quid pro quo.* Apollo did whatever it is that gods do to create seers, oracles, and prophets out of mere mortals. But then, scandalously, Cassandra reneged. She refused the overtures of a god.

Apollo was incensed. But he couldn't withdraw the gift of prophecy, because, after all, he was a god. (Whatever else you might say about them, gods keep their promises.) Instead, he condemned her to a cruel and ingenious fate: that no one would believe her prophecies. (What I'm recounting here is largely from Aeschylus's play *Agamemnon.*) Cassandra prophesies to her own people the fall of Troy. Nobody pays attention. She predicts the death of the leading Greek invader, Agamemnon. Nobody pays attention. She even foresees her own early death, and still no one pays attention. They didn't want to hear. They made fun of her. They called her — Greeks and Trojans alike — "the lady of many sorrows." Today perhaps they would dismiss her as a "prophet of doom and gloom."

There's a nice moment when she can't understand how it is that these prophecies of impending catastrophe — some of which, if believed, could be prevented — were being ignored. She says to the Greeks, "How is it you don't understand me? Your tongue I know only too well." But the problem wasn't her pronunciation of Greek. The answer (I'm paraphrasing) was, "You see, it's like this. Even the Delphic Oracle sometimes makes mistakes. Sometimes its prophecies are ambiguous. We can't be sure. And if we can't be sure about Delphi, we certainly can't be sure about you." That's the closest she gets to a substantive response.

The story was the same with the Trojans: "I prophesied to my countrymen," she says, "all their disasters." But they ignored her clairvoyances and were destroyed. Soon, so was she.

The resistance to dire prophecy that Cassandra experienced can be recognized today. If we're faced with an ominous prediction involving powerful forces that may not be readily influenced, we have a natural tendency to reject or ignore the prophecy. Mitigating or circumventing the danger might take time, effort, money, courage. It might require us to alter the priorities of our lives. And not every prediction of disaster, even among those made by scientists, is fulfilled: Most animal life in the oceans did not perish due to insecticides; despite Ethiopia and the Sahel, worldwide famine has not been a hallmark of the 1980s; food production in South Asia was not drastically affected by the 1991 Kuwaiti oil well fires; supersonic transports do not threaten the ozone layer — although all these predictions had been made by serious scientists. So when faced with a new and uncomfortable prediction, we might be tempted to say: "Improbable." "Doom and Gloom." "We've never experienced anything remotely like it." "Trying to frighten everyone." "Bad for public morale."

What's more, if the factors precipitating the anticipated catastrophe are long-standing, then the prediction itself is an indirect or unspoken rebuke. Why have we, ordinary citizens, permitted this peril to develop? Shouldn't we have informed ourselves about it earlier? Don't we ourselves bear complicity, since we didn't take steps to insure that government leaders eliminated the threat? And since these are uncomfortable ruminations — that our own inattention and inaction may have put us and our loved ones in danger — there is a natural, if maladaptive, tendency to reject the whole business. It will need much better evidence, we say, before we can take it seriously. There is a temptation to minimize, dismiss, forget. Psychiatrists are fully aware of this temptation. They call it "denial." As the lyrics of an old rock song go: "Denial ain't just a river in Egypt."

The stories of Croesus and Cassandra represent the two extremes of policy response to predictions of deadly peril — Croesus himself representing one pole of credulous, uncritical acceptance (usually of the assurance that all is well), propelled by greed or other character flaws; and the Greek and Trojan response to Cassandra representing the pole of stolid, immobile rejection of the possibility of danger. The job of the policymaker is to steer a prudent course between these two shoals.

Suppose a group of scientists claims that a major environmental catastrophe is looming. Suppose further that what is required to prevent or mitigate the catastrophe is expensive: expensive in fiscal and intellectual resources, but also in challenging our way of thinking — that is, politically expensive. At what point do the policymakers have to take the scientific prophets seriously? There are ways to assess the validity of the modern prophecies — because in the methods of science, there is an error-correcting procedure, a set of rules that have repeatedly worked well, sometimes called the scientific method. There are a number of tenets (I've outlined some of them in my book *The Demon-Haunted World*): Arguments from authority carry little weight ("Because I said so" isn't good enough); quantitative prediction is an extremely good way to sift useful ideas from nonsense; the methods of analysis must yield other results fully consistent with what else we know about the Universe; vigorous debate is a healthy sign; the same conclusions have to be drawn independently by competent competing scientific groups for an idea to be taken seriously; and so on. There are ways for policymakers to decide, to find a safe middle path between precipitate action and impassivity. It takes some emotional discipline, though, and most of all an aware and scientifically literate citizenry — able to judge for themselves how dire the dangers are.

Chapter 10

A PIECE OF THE SKY IS MISSING

> [T]his goodly frame, the earth, seems to me a sterile promontory; this most excellent canopy, the air, look you, this brave o'erhanging firmament, this majestical roof fretted with golden fire, why, it appears no other thing to me than a foul and pestilent congregation of vapors.
>
> WILLIAM SHAKESPEARE,
> *Hamlet*, II, ii, 308 (1600–1601)

I'd always wanted a toy electric train. But it wasn't until I was 10 that my parents could afford to buy me one. What they got me, secondhand but in good condition, wasn't one of those bantamweight, finger-long, miniature scale models you see today, but a real clunker. The locomotive alone must have weighed five pounds. There was also a coal tender, a passenger car, and a caboose. The all-metal interlocking tracks came in three varieties: straight, curved, and one beautifully crossed mutation that permitted the construction of a figure-eight railway. I saved up to buy a green plastic tunnel, so I could see the engine, its headlight dispelling the darkness, triumphantly chugging through.

My memories of those happy times are suffused with a smell — not unpleasant, faintly sweet, and always emanating from the transformer, a big black metal box with a sliding red lever that controlled the speed of the train. If you had asked me to describe its function, I suppose I would have said that it converted the kind of electricity in the walls of our apartment to the kind of electricity that the locomotive needed. Only much later did I learn that the smell was made by a particular chemical — generated by the electricity as it passed through air — and that the chemical had a name: ozone.

The air all around us, the stuff we breathe, is made of about 20 percent oxygen — not the atom, symbolized as O, but the molecule, symbolized as O_2, meaning two oxygen atoms chemically bound together. This molecular oxygen is what makes us go. We breathe it in, combine it with food, and extract energy. Ozone is a much rarer way in which oxygen atoms combine. It is symbolized as O_3, meaning three oxygen atoms chemically bound together.

My transformer had an imperfection. A tiny electric spark had been sputtering away, breaking the bonds of oxygen molecules as they happened by:

$$O_2 + \text{energy} \rightarrow O + O$$

(The arrow means *is changed into*.) But solitary oxygen atoms (O) are unhappy, chemically reactive, anxious to combine with adjacent molecules — and they do:

$$O + O_2 + M \rightarrow O_3 + M$$

Here, M stands for any third molecule; it doesn't get used up in the reaction but is required to help it along. M is a catalyst. There are plenty of M molecules around, chiefly molecular nitrogen.

That's what was going on in my transformer to make ozone. It also goes on in automobile engines and in the fires of industry, producing reactive ozone down here near the ground, contributing to smog and industrial pollution. It doesn't smell so sweet to me anymore. The biggest ozone danger is not too much of it down here, but too little of it up there.

It was all done responsibly, carefully, with concern for the environment. By the 1920s, refrigerators were widely perceived to be a good thing. For reasons of convenience, public health, the ability of producers of fruit, vegetables, and milk products to market at sizable distances, and tasty meals combined, everyone wanted to have one. (No more lugging blocks of ice; what could be bad about that?) But the working fluid, whose heating and cooling provided the refrigeration, was either ammonia or sulfur dioxide — poisonous and evil-smelling gases. A leak was very ugly. A substitute was badly needed — one that was liquid under the right conditions, that would circulate inside the refrigerator but would not hurt anything if the refrigerator leaked or was converted into scrap metal. For these purposes it would be nice to find a material that was also neither poisonous nor flammable, that doesn't corrode, burn your eyes, attract bugs, or even bother the cat. But in all of Nature, no such material seemed to exist.

So chemists in the United States and Weimar and Nazi Germany invented a class of molecules that had never existed on Earth before. They called them chlorofluorocarbons (CFCs), made up of one or more carbon atoms to which are attached some chlorine and/or fluorine atoms. Here's one:

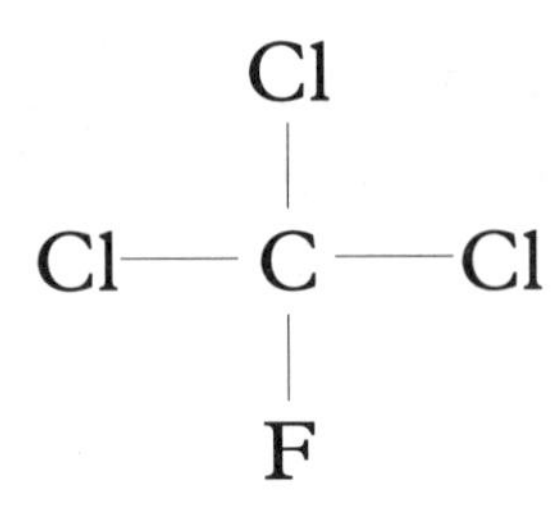

(C for carbon, Cl for chlorine, F for fluorine.) They were wildly successful, far exceeding the expectations of their inventors. Not only did they become the chief working fluid in refrigerators, but in air conditioners as well. They found widespread applications in aerosol spray cans, insulating foam, and industrial solvents and cleansing agents (especially in the microelectronics industry). The most famous brand name is Freon, a trademark of DuPont. It was used for decades and no harm seemed ever to come from it. Safe as safe could be, everyone figured. That's why, after a while, a surprising amount of what we took for granted in industrial chemistry depended on CFCs.

By the early 1970s a million tons of the stuff were manufactured every year. So, it's the early 1970s, let's say, and you're standing in your bathroom, spraying under your arms. The CFC aerosol comes out in a fine deodorant-carrying mist. The propellant CFC molecules don't stick to you. They bounce off into the air, swirl near the mirror, careen off the walls. Eventually, some of them trickle out the window or under the door and, as time passes — it may take days or weeks — they find themselves in the great outdoors. The CFCs bump into other molecules in the air, off buildings and telephone poles, and, carried up by convection currents and by the global atmospheric circulation, are swept around the planet. With very few exceptions, they do not fall apart and do not chemically combine with any of the other molecules they encounter. They're practically inert. After a few years, they find themselves in the high atmosphere.

Ozone is naturally formed up there at an altitude of around 25 kilometers (15 miles). Ultraviolet light (UV) from the Sun — corresponding to the spark in my imperfectly insulated electric-train transformer — breaks O_2 molecules down into O atoms. They recombine and reform ozone, just as in my transformer.

A CFC molecule survives at those altitudes on the average for a century before the UV makes it give up its chlorine. Chlorine is a catalyst that destroys ozone molecules but is not destroyed itself. It takes a couple of years before the chlorine is carried back into the lower atmosphere and washed out in rainwater. In that time, a chlorine atom may preside over the destruction of 100,000 ozone molecules.

The reaction goes like this:

$$O_2 + \text{UV light} \rightarrow 2O$$
$$2Cl\ [\text{from CFCs}] + 2O_3 \rightarrow 2ClO + 2O_2$$
$$2ClO + 2O \rightarrow 2Cl\ [\text{regenerating the Cl}] + 2O_2$$

So the net result is:

$$2O_3 \rightarrow 3O_2$$

Two ozone molecules have been destroyed; three oxygen molecules have been generated; and the chlorine atoms are available to do further mischief.

So what? Who cares? Some invisible molecules, somewhere high up in the sky, are being destroyed by some other invisible molecules manufactured down here on Earth. Why should we worry about that?

Because ozone is our shield against ultraviolet light from the Sun. If all the ozone in the upper air were brought down to the temperature and pressure around you at this moment, the layer would be only three millimeters thick — about the height of the cuticle of your little finger if you're not fastidiously manicured. It's not very much ozone. But that ozone is all that stands between us and the fierce and searing longwave UV from the Sun.

The UV danger we often hear about is skin cancer. Light-skinned people are especially vulnerable; dark-skinned people have a generous supply of melanin to protect them. (Suntanning is an adaptation whereby whites develop more protective melanin when exposed to UV.) There seems to be some remote cosmic justice in light-skinned people inventing CFCs, which then give skin cancer preferentially to light-skinned people, while dark-skinned people, having had little to do with this wonderful invention, are naturally protected. There are ten times more malignant skin cancers reported today than in the 1950s. While part of this increase may be due to better reporting, ozone loss and increased UV exposure seem implicated. If things were to get much worse, light-skinned people might be required to use special protective clothing during routine excursions out-of-doors, at least at highish altitudes and latitudes.

But increased skin cancer, while a direct consequence of enhanced UV, and threatening millions of deaths, is not the worst of it. Nor is the increased rate of eye cataracts. More serious is the fact that UV injures the immune system — the body's machinery for fighting disease — but, again, only for people who go out unprotected into the sunlight. Yet, as serious as *this* seems, the real danger lies elsewhere.

When exposed to ultraviolet light, the organic molecules that constitute all life on Earth fall apart or make unhealthy chemical attachments. The most prevalent beings that inhabit the oceans are tiny one-celled plants that float near the surface of the water — the phytoplankton. They can't hide from the UV by diving deep because they make a living through harvesting sunlight. They live from hand to mouth (a metaphor only — they have neither hands nor mouths). Experiments show that even a moderate increase in UV harms the one-celled plants common in the Antarctic Ocean and elsewhere. Larger increases can be expected to cause profound distress and, eventually, massive deaths.

Preliminary measurements of populations of these microscopic plants in Antarctic waters show that there has recently been a striking decline — up to 25 percent — near the ocean's surface. Phytoplankton, because they're so small, lack the tough UV-absorbing skins of animals and higher plants. (In addition to a set of cascading consequences in the oceanic food chain, the deaths of phytoplankton eliminates their ability to extract carbon dioxide from the atmosphere — and thereby adds to global warming. This is one of several ways in which the thinning of the ozone layer and the heating of the Earth are connected — even though they are fundamentally very different questions. The main action for ozone depletion occurs in the ultraviolet; for global warming, in the visible and infrared.)

But if increasing UV falls on the oceans, the damage is not restricted to these little plants — because they are the food of one-celled animals (the zooplankton), who are eaten in turn by little shrimplike crustaceans (like those in my glass world number 4210 — the krill), who are eaten by small fish, who are eaten by large fish, who are eaten by dolphins, whales, and people. The destruction of the little plants at the base of the food chain causes the entire chain to collapse. There are many such food chains, on land as in water, and all seem vulnerable to disruption by UV. For example, the bacteria in the roots of rice plants that grab nitrogen from the air are UV-sensitive. Increasing UV may threaten crops and possibly even compromise the human food supply. Laboratory studies of crops at mid-latitudes show that many are injured by increases in the near-ultraviolet light that is let through as the ozone layer thins.

In permitting the ozone layer to be destroyed and the intensity of UV at the Earth's surface to increase, we are posing challenges of unknown but worrisome severity to the fabric of life on our planet. We are ignorant about the complex mutual dependencies of the beings on Earth, and what the sequential consequences will be if we wipe out some especially vulnerable microbes on which larger organisms depend. We are tugging at a planetwide biological tapestry and do not know whether one thread only will come out in our hands, or whether the whole tapestry will unravel before us.

No one believes that the entire ozone layer is in imminent danger of disappearing. We will not — even if we remain wholly obdurate about acknowledging our danger — be reduced to the antiseptic circumstance of the Martian surface, pummeled by unfiltered solar UV. But even a worldwide reduction in the amount of ozone by 10 percent — and many scientists think that's what the *present* dose of CFCs in the atmosphere will eventually bring about — looks to be very dangerous.

In 1974, F. Sherwood Rowland and Mario Molina of the Irvine campus of the University of California first warned that CFCs — some million tons per year were being injected into the stratosphere — might seriously damage the ozone layer. Subsequent experiments and calculations by scientists all over the world have supported their findings. At first certain confirmatory calculations suggested the effect was there, but would be less serious than Rowland and Molina proposed; other calculations suggested it would be more serious. This is a common circumstance for a new scientific finding, as other scientists try to find out how robust the new discovery is. But the calculations settled down more or less where Rowland and Molina said they would. (And in 1995 they would share the Nobel Prize in Chemistry for this work.)

DuPont, which sold CFCs to the tune of $600 million a year, took out ads in newspapers and scientific journals, and testified before Congressional committees that the danger of CFCs to the ozone layer was unproved, had been greatly exaggerated, or was based on faulty scientific reasoning. Its ads compared "theorists and some legislators," who were for banning CFCs in aerosols, with "researchers and the aerosol industry," who were for temporizing. It argued that "other chemicals . . . are primarily responsible" and warned about "businesses destroyed by premature legislative action." It claimed a "lack of evidence" on the issue and promised to begin three years of research, after which they might do something. A powerful and profitable corporation was not about to risk hundreds of millions of dollars a year on the mere say-so of a few photochemists. When the theory was proven beyond the shadow of a doubt, they in effect said, that would be soon enough to consider making changes. Sometimes they seemed to be arguing that CFC manufacture would be halted as soon as the ozone layer was irretrievably damaged. But by then there might be no customers.

Once CFCs are in the atmosphere, there is no way to scrub them out (or to pump ozone from down here, where it's a pollutant, to up there, where it's needed). The effects of CFCs, once introduced into the air, will persist for about a century. Thus

Sherwood Rowland, other scientists, and the Washington-based Natural Resources Defense Council urged the banning of CFCs. By 1978, CFC propellants in aerosol spray cans were made illegal in the United States, Canada, Norway, and Sweden. But most world CFC production did not go into spray cans. Public concern was temporarily mollified, attention drifted elsewhere, and the CFC content of the air continued to rise. The amount of chlorine in the atmosphere reached twice what it was when Rowland and Molina sounded the alarm and five times what it was in 1950.

For years, the British Antarctic Survey, a team of scientists stationed at Halley Bay in the southernmost continent, had been measuring the ozone layer high overhead. In 1985 they announced the disconcerting news that the springtime ozone had diminished to nearly half what they had measured a few years before. The discovery was confirmed by a NASA satellite. Two-thirds of the springtime ozone over Antarctica is now missing. There's a hole in the Antarctic ozone layer. It has shown up every spring since the late 1970s. While it heals itself in winter, the hole seems to last longer each spring. No scientist had predicted it.

Naturally, the hole led to more calls for

a ban on CFCs (as did the discovery that CFCs add to the global warming caused by the carbon dioxide greenhouse effect). But industry officials seemed to have difficulty focusing on the nature of the problem. Richard C. Barnett, chairman of the Alliance for a Responsible CFC Policy — formed by CFC manufacturers — complained: "The rapid, complete shutdown of CFCs that some people are calling for would have horrendous consequences. Some industries would have to shut down because they cannot get alternative products — the cure could kill the patient." But the patient is not "some industries"; the patient might be life on Earth.

The Chemical Manufacturers Association believed that the Antarctic hole "is highly unlikely to have global significance. . . . Even in the other most similar region of the world, the Arctic, the meteorology effectively precludes a similar situation."

More recently, enhanced levels of reactive chlorine have been found *in* the ozone hole, helping to establish the CFC connection. And measurements near the North Pole suggest that an ozone hole *is* developing over the Arctic as well. A 1996 study called "Satellite confirmation of the dominance of chlorofluorocarbons in the global stratospheric chlorine budget" draws the unusually strong conclusion (for a scientific paper) that CFCs are implicated in ozone depletion "beyond reasonable doubt." The role of chlorine from volcanos and sea spray — advocated by some right-wing radio commentators — accounts at most for 5 percent of the destroyed ozone.

At northern midlatitudes, where most people on Earth live, the amount of ozone seems to have been steadily declining at least since 1969. There are fluctuations, of course, and volcanic aerosols in the stratosphere work to decrease ozone levels for a year or two before they settle out. But to find (according to the World Meteorological Organization) 30 percent relative depletions over northern midlatitudes for some months of each year, and 45 percent in some areas, is cause for alarm. You don't need many consecutive years like that before it's likely that the life underneath the thinning ozone layer is getting into trouble.

Berkeley, California, banned the white CFC-blown-foam insulation used to keep fast foods warm. McDonald's pledged replacement of the most damaging CFCs in its packaging. Facing the threat of government regulations and consumer boycotts, DuPont finally announced in 1988, 14 years after the CFC danger had been identified, that it would phase out the manufacture of CFCs — but not to be completed until the year 2000. Other American manufacturers did not then promise even that. The United States, though, accounted for only 30 percent of worldwide CFC production. Clearly, since the long-term threat to the ozone layer is global, the solution would have to be global as well.

In September 1987, many of the nations that produce and use CFCs met in Montreal to consider a possible agreement to limit CFC use. At first, Britain, Italy, and France, influenced by their powerful chemical industries (and France by its perfume industry), participated in the discussions only reluctantly. (They feared that DuPont had a substitute up its sleeve that it had been preparing all the time it was stonewalling about CFCs. The United States was pushing a ban on CFCs, they worried, in order to increase the global competitiveness of one of its major corporations.) Such nations as South Korea were altogether absent. The Chinese delegation did not sign the treaty. Interior Secretary Donald Hodel, a conservative Reagan appointee averse to government controls, reportedly suggested that, instead of limiting CFC production, we all wear sunglasses and hats. This option is unavailable to the microorganisms at the base of the food chains that sustain life on Earth. The United States signed the Montreal Protocol despite this advice. That this occurred during the antienvironmental spasm of the late Reagan Administration was truly unexpected (unless, of course, the fear of DuPont's European competitors is true.) In the United States alone, 90 million vehicle air conditioners and 100 million refrigerators would have to be replaced. This represented a considerable sacrifice to preserve the environment. Substantial credit must be given to Ambassador Richard Benedick, who led the U.S. delegation to Montreal, and to British Prime Minister Margaret Thatcher, who, trained in chemistry, understood the issue.

The Montreal Protocol has now been strengthened still further by amendment agreements signed in London and Copenhagen. As of this writing, 156 nations, including the republics of the former Soviet Union, China, South Korea, and India have signed the treaty. (Although some nations ask why, when Japan and the West have benefitted from CFCs, they must forgo refrigerators and air conditioners, just when their industries are hitting stride. It is a fair question, but a very narrow one.) A total phaseout of CFCs was agreed to by the year 2000, and then amended to 1996. China, whose CFC consumption was rising by 20 percent per year through the 1980s, agreed to cut its reliance on CFCs and not avail itself of a ten-year grace period that the agreement permitted. DuPont has become a leader in cutting back on CFCs, and has committed itself to a faster phaseout than many nations have. The amount of CFCs in the atmosphere is measurably declining. The trouble is, we will have to stop producing *all* CFCs and then wait a century before the atmosphere cleans itself up. The longer we dawdle, the more nations that hold out, the greater the danger.

Clearly, the problem is solved if a cheaper and more effective CFC substitute can be found that does not injure us or the environment. But what if there is no such substitute? What if the best substitute is more expensive than CFCs? Who pays for the research, and who makes up the price difference — the consumer, the government, or the chemical industry that got us into (and profited by) this mess? Do the industrialized nations who benefited from CFC technology give significant aid to the emerging industrialized states who have not? What if we need 20 years to be sure the substitute doesn't cause cancer? What about the UV now pouring down on the Antarctic Ocean? What about all the newly manufactured CFCs rising toward the ozone layer between now and whenever the stuff is completely banned?

A substitute — or better, a stopgap measure — has been found. CFCs are temporarily being replaced by HCFCs, similar molecules but involving hydrogen atoms. For example:

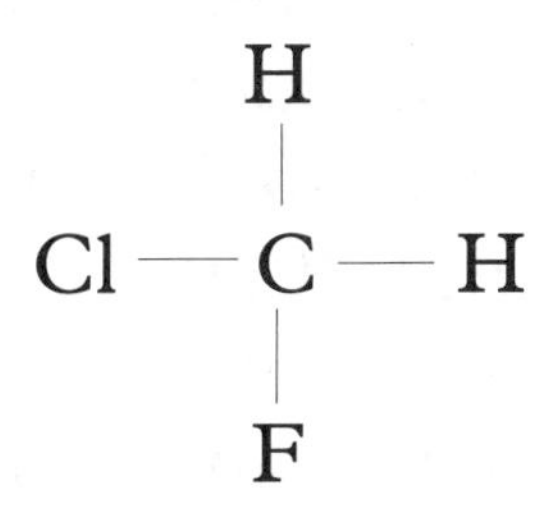

They still cause some damage to the ozone layer, but much less; they are, like CFCs, a significant contributor to global warming; and, especially during start-up, they are more expensive. But they do address the most immediate need, protecting the ozone layer. HCFCs were developed by DuPont, but — the company swears — only *after* the discoveries at Halley Bay.

Bromine is, atom-for-atom, at least 40 times more effective than chlorine in destroying stratospheric ozone. Fortunately, it is much rarer than chlorine. Bromine is released to the air in halons used in fire extinguishers, and methyl bromide,

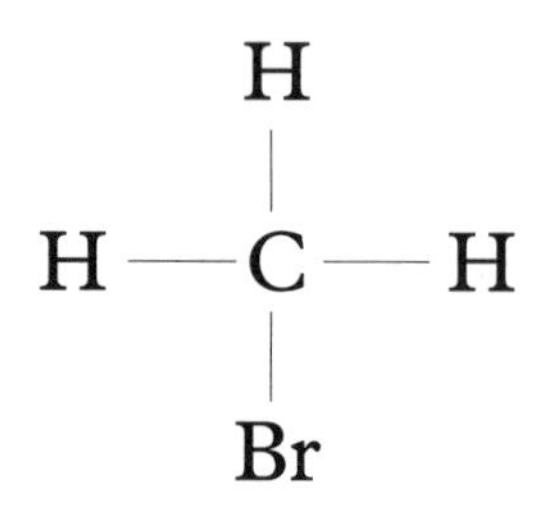

used to fumigate soil and stored grain. In 1994–96, the industrial nations agreed to phase out the production of these materials, capping them by 1996, but not totally phasing them out until 2030. Because there are as yet no replacements for some halons, there may be a temptation to keep on using them — ban or no ban. Meanwhile, a major technological issue is finding a superior long-term solution to overtake HCFCs. This might involve another brilliant synthesis of a new molecule, but perhaps will go in other directions — for example, acoustic refrigerators that have no circulating fluid carrying subtle dangers. Here is an opportunity for creative invention. Both the financial rewards and the long-term benefit for the species and the planet are high. I'd like to see the enormous technical skill at the nuclear weapons laboratories, now increasingly moribund because of the end of the Cold War, turned to such worthy pursuits. I'd like to see generous grants and irresistible prizes offered to invent effective, convenient, safe, and reasonably inexpensive new modes for air conditioners and refrigerators — that are appropriate for local manufacture in developing nations.

The Montreal Protocol is important for the magnitude of the changes agreed to but especially for their direction. Perhaps most surprising, a ban on CFCs was agreed to when it was unclear that there would be a feasible alternative. The Montreal conference was sponsored by the United Nations Environment Programme, whose director, Mostafa K. Tolba, described it as "the first truly global treaty that offers protection to every single human being."

It is an encouragement that we can recognize new and unexpected dangers, that the human species can come together working on behalf of all of us on such an issue, that rich nations might be willing to bear a fair share of the cost, and that corporations with much to lose can be made not only to change their minds, but to see in such a crisis new entrepreneurial opportunities. The CFC ban provides what in mathematics is known as an existence theorem — a demonstration that something that might, for all you know, be impossible can actually be accomplished. It is a reason for cautious optimism.

Chlorine seems to have peaked at about four chlorine atoms for every billion other molecules in the stratosphere. The amount is now decreasing. But at least partly because of bromine, the ozone layer is not predicted to heal itself soon.

Clearly, it's too early to wholly relax on protecting the ozone layer. We need to make sure that the manufacture of these materials is almost entirely stopped all over the world. We need greatly enhanced research to find safe substitutes. We need comprehensive monitoring (from ground stations, airplanes, and satellites in orbit) of the ozone layer all over the globe* at least

*The National Aeronautics and Space Administration and the National Oceanic and Atmospheric Administration have played heroic roles in acquiring data about the depleting ozone layer and its causes. (The *Nimbus-7* satellite, for example, found an increase of the most dangerous UV wavelengths reaching the Earth's surface of 10 percent a decade for southern Chile and Argentina and about half that at northern midlatitudes, as conscientiously as we would watch over a loved one suffering from heart palpitations. We need to know by how much the ozone layer is further stressed by occasional volcanic explosions, or continued global warming, or the introduction of some new chemical into the world atmosphere.

Starting shortly after the Montreal Protocol, stratospheric chlorine levels have declined. Starting in 1994 the stratospheric chlorine and bromine levels (taken together) have declined. If bromine levels also decline, the ozone layer should, it is estimated, begin a long-term recovery by the turn of the century. Had no CFC controls been instituted until 2010, stratospheric chlorine would have climbed to levels three times higher than today's, the Antarctic ozone hole would have persisted until the mid-

---

where most people on Earth live.) A new NASA satellite program called Mission to Planet Earth will continue monitoring ozone and related atmospheric phenomena on an ambitious scale for a decade or more. Meanwhile, Russia, Japan, the constituent members of the European Space Agency, and others are weighing in with their own programs and their own spacecraft. By these criteria also, the human species is taking the threat of ozone depletion seriously.

twenty-second century, and springtime ozone depletion in the northern midlatitudes might have reached well above 30 percent, a whopping value — according to Rowland's Irvine colleague Michael Prather.

In the United States there is still resistance from the air conditioning and refrigerator industries, from extreme "conservatives," and from Republican members of Congress. Tom DeLay, the Republican House majority whip, was in 1996 of the opinion that "the science underlying the CFC ban is debatable," and that the Montreal Protocol is "the result of a media scare." John Doolittle, another House Republican, insisted that the causal link connecting ozone depletion with CFCs is "still very much open to debate." In response to a reporter who reminded him of the critical, skeptical peer review by experts that papers establishing this link had been subjected to, Doolittle replied, "I'm not going to get involved in peer-review mumbo-jumbo." It might be better for the country if he did. Peer review is in fact the great mumbo-jumbo detector. The Nobel Committee's judgment was different. In conferring the Prize on Rowland and Molina — whose names should be known to every schoolchild — it commended them for having "contributed to our salvation from a global environmental problem that could have catastrophic consequences." It's hard to understand how "conservatives" could oppose safeguarding the environment that all of us — including conservatives and their children — depend on for our very lives. What exactly is it conservatives are conserving?

The central elements of the ozone story are like many other environmental threats: We pour some substance into the atmosphere (or prepare to do so). Somehow we do not thoroughly examine its environmental impact — because examination would be expensive, or would delay production and cut into profits; or because those in charge do not want to hear counterarguments; or because the best scientific talent has not been brought to bear on the issue; or simply because we're human and fallible and have missed something. Then, suddenly, we are face-to-face with a wholly unexpected danger of worldwide dimensions that may have its most ominous consequences decades or centuries from now. The problem cannot be solved locally, or in the short term.

In all these cases, the lesson is clear: We are not always smart or wise enough to foresee all the consequences of our actions. The invention of CFCs was a brilliant achievement. But as smart as those chemists were, they weren't smart enough. Precisely because CFCs are so inert, they survived long enough to reach the ozone layer. The world is complicated. The air is thin. Nature is subtle. Our capacity to cause harm is great. We must be much more careful and much less forgiving about polluting our fragile atmosphere.

We must develop higher standards of planetary hygiene and significantly greater scientific resources for monitoring and understanding the world. And we must begin to think and act not merely in terms of our nation and generation (much less the profits of a particular industry) but in terms of the entire vulnerable planet Earth and the generations of children to come.

The hole in the ozone layer is a kind of skywriting. At first it seemed to spell out our continuing complacency before a witch's brew of deadly perils. But perhaps it really tells of a newfound talent to work together to protect the global environment. The Montreal Protocol and its amendments represent a triumph and a glory for the human species.

Chapter 11

AMBUSH: THE WARMING OF THE WORLD

They set an ambush for their own lives.
Proverbs 1:18

Three hundred million years ago the Earth was covered by vast swamps. When the ferns, horsetails, and club mosses died, they were buried in muck. Ages passed; the remains were carried down underground and there transformed by slow stages into a hard organic solid that we call coal. In other locales and epochs, immense numbers of one-celled plants and animals died, sank to the sea floor, and were covered by sediment. Simmering for ages, their remains were, by imperceptible steps, converted to buried organic liquids and gases that we call petroleum and natural gas. (Some additional natural gas may be primordial — not of biological origin but incorporated into the Earth during its formation.) After humans evolved there were occasional early encounters with these strange materials when they were carried to the Earth's surface. Seepage of oil and gas and their ignition by lightning is thought to be the origin of the "eternal flame" central to the fire worshiping religions of ancient Persia. Marco Polo was widely disbelieved when he told the European experts of his day the preposterous story that in China a black rock was mined that burned when lit.

Eventually, the Europeans recognized that these readily transported, energy-rich materials could be useful. They were much better than wood. You could warm your house with them, stoke a furnace, drive a steam engine, generate electricity, power industry, and make trains, cars, ships, and planes go. And there were potent military applications. So we learned to dig the coal out of the Earth and drive deep holes into the ground so the deeply buried gas and oil, compressed by the overburden of rock, could come shooting out to the surface. Eventually, these substances came to dominate the economy. They have provided the propulsion for our global technological civilization. It is no exaggeration to say that in a sense they run the world. As always, there is a price to pay.

Coal, oil, and gas are called fossil fuels, because they are mostly made of the fossil remains of beings from long ago. The chemical energy within them is a kind of stored sunlight originally accumulated by ancient plants. Our civilization runs by burning the remains of humble creatures who inhabited the Earth hundreds of millions of years before the first humans came on the scene. Like some ghastly cannibal cult, we subsist on the dead bodies of our ancestors and distant relatives.

If we think back to the time when our only fuel was wood, we gain some appreciation of the benefits fossil fuels have brought. They have also created vast global industries, with immense financial and political power — not just the oil, gas, and coal conglomerates, but also subsidiary industries wholly (autos, airplanes) or partly (chemicals, fertilizers, agriculture) dependent on them. This dependence means that nations will go to extreme lengths to preserve their sources of supply. Fossil fuels were important factors in the conduct of World Wars I and II. Japanese aggression at the start of World War II was explained and justified on the grounds that she was obliged to safeguard her sources of oil. As, for example, the 1991 Persian Gulf War reminds us, the political and military importance of fossil fuels remains high.

About 30 percent of all U.S. oil imports comes from the Persian Gulf. In some months, more than half of U.S. oil is imported. Oil constitutes more than half of all U.S. balance of payments deficits. The U.S. spends over a billion dollars a week in oil imports from abroad. Japan's oil import bill is about the same. China — with burgeoning consumer demand for autos — may reach the same level early in the twenty-first century. Similar numbers apply to Western Europe. Economists spin scenarios in which increases in oil prices induce inflation, higher interest rates, diminished investment in new industry, fewer jobs, and economic recession. It may not happen, but it is a possible consequence of our addiction to oil. Oil drives nations into policies they might otherwise find unprincipled or foolhardy. Consider, for example, the following (1990) comment from the syndicated columnist Jack Anderson, expressing a widely held opinion: "As unpopular as the notion is, the United States must continue to be policeman for the globe. On a purely selfish level, Americans need what the world has — oil being the pre-eminent need." According to Bob Dole, the Senate minority leader at the time, the Persian Gulf War — which put over 200,000 young American men and women at risk — was undertaken "for one reason only: O-I-L."

As I write, the nominal cost of crude oil is almost $20 a barrel, while the world's authenticated or "proven" petroleum reserves are almost a trillion barrels. Twenty trillion dollars is four times the U.S. national debt, the largest in the world. Black gold, indeed.

The global production of petroleum is about 20 billion barrels a year, so each year we use up about 2 percent of the proven reserves. You might think we're going to run out pretty soon, maybe in the next 50 years. But we keep finding new reserves. Previous predictions that we would run out of petroleum by such-and-such a date have always proved baseless. There is a finite amount of oil, gas, and coal in the world, it's true. There were only so many of those ancient organisms that contributed their bodies for our comfort and convenience. But it seems unlikely we will run out of fossil fuels soon. The only problem is, it's more and more expensive to find new and unexploited reserves, the world economy can go into fibrillation if oil prices are made to change quickly, and countries go to war to get the stuff. Also, of course, there's the environmental cost.

The price we pay for fossil fuels is measured not just in dollars. The "satanic mills" of England in the early years of the Industrial Revolution polluted the air and caused an epidemic of respiratory disease. The "pea soup" fogs of London, so familiar to us from dramatizations of Holmes and Watson, Jekyll and Hyde, and Jack the Ripper and his victims, were deadly domestic and industrial pollution — largely from burning coal. Today, automobiles add their exhaust fumes, and our cities are plagued by smog — which affects the health, happiness, and productivity of the very people generating the pollutants. We also know about acid rain and the ecological turmoil caused by oil spills. But the prevailing opinion has been that these penalties to health and environment were more than balanced by the benefits that fossil fuels bring.

Now, though, the governments and peoples of the Earth are gradually becoming aware of yet another dangerous consequence of the burning of fossil fuels: If I burn a piece of coal or a gallon of petroleum or a cubic foot of natural gas, I'm combining the carbon in the fossil fuel with the oxygen in the air. This chemical reaction releases energy locked away for perhaps 200 million years. But in combining a carbon atom, C, with an oxygen molecule, O_2, I also synthesize a molecule of carbon dioxide, CO_2.

$$C + O_2 \rightarrow CO_2$$

And CO_2 is a greenhouse gas.

What determines the average temperature of the Earth, the planetary climate? The amount of heat trickling up from the center of the Earth is negligible compared with the amount falling down on the Earth's surface from the Sun. Indeed, if the Sun were turned off, the temperature of the Earth would fall so far that the air would freeze solid, and the planet would be covered with a layer of nitrogen and oxygen snow 10-meters (30-feet) thick. Well, we know how much sunlight is falling on the Earth and warming it. Can't we calculate what the average temperature of the Earth's surface ought to be? This is an easy calculation — taught in elementary astronomy and meteorology courses, another example of the power and beauty of quantification.

The amount of sunlight absorbed by the Earth has to equal on average the amount of energy radiated back to space. We don't ordinarily think of the Earth as radiating into space, and when we fly over it at night we don't see it glowing in the dark (except for cities). But that's because we're looking in ordinary visible light, the kind to which our eyes are sensitive. If we were to look beyond red light into what's called the thermal infrared part of the spectrum — at 20 times the wavelength of yellow light, for example — we would see the Earth glowing in its own eerie, cool infrared light, more in the Sahara than Antarctica, more in daytime than at night. This is not sunlight reflected off the Earth, but the planet's own body heat. The more energy coming in from the Sun, the more the Earth radiates back to space. The hotter the Earth, the more it glows in the dark.

What's coming in to warm the Earth depends on how bright the Sun is and how reflective the Earth is. (Whatever isn't reflected back into space is absorbed by the ground, the clouds, and the air. If the Earth were perfectly shiny and reflective, the sunlight falling on it wouldn't warm it up at all.) The reflected sunlight, of course, is mainly in the visible part of the spectrum. So set the input (which depends on how much sunlight the Earth absorbs) equal to the output (which depends on the temperature of the Earth), balance the two sides of the equation, and out comes the predicted temperature of the Earth. A cinch! Couldn't be easier! You calculate it, and what's the answer?

Our calculation tells us that the average temperature of the Earth should be about 20° Celsius below the freezing point of water. The oceans ought to be blocks of ice and we all ought to be frozen stiff. The Earth should be inhospitable to almost all forms of life. What's wrong with the calculation? Did we make a mistake?

We didn't exactly make a mistake in the calculation. We just left something out: the greenhouse effect. We implicitly assumed that the Earth had no atmosphere. While the air is transparent at ordinary visible wavelengths (except for places like Denver and Los Angeles), it's much more opaque in the thermal infrared part of the spectrum, where the Earth likes to radiate to space. And that makes all the difference in the world. Some of the gases in the air in front of us — carbon dioxide, water vapor, some oxides of nitrogen, methane, chlorofluorocarbons — happen to absorb strongly in the infrared, even though they are completely transparent in the visible. If you put a layer of this stuff above the surface of the Earth, the sunlight still gets in. But when the surface tries to radiate back to space, the way is impeded by this blanket of infrared absorbing gases. It's transparent in the visible, semi-opaque in the infrared. As a result the Earth has to warm up some, to achieve the equilibrium between the sunlight coming in and the infrared radiation emitted out. If you calculate how opaque these gases are in the infrared, how much of the Earth's body heat they intercept, you come out with the right answer. You find that on average — averaged over seasons, latitude, and time of day — the Earth's surface must be some 13°C above zero. This is why the oceans don't freeze, why the climate is congenial for our species and our civilization.

Our lives depend on a delicate balance of invisible gases that are minor components of the Earth's atmosphere. A little greenhouse effect is a good thing. But if you add more greenhouse gases — as we have been doing since the beginning of the Industrial Revolution — you absorb more infrared radiation. You make that blanket thicker. You warm the Earth further.

For the public and policymakers, all this may seem a little abstract — invisible gases, infrared blankets, calculations by physicists.

If difficult decisions on spending money are to be made, don't we need a little more evidence that there really *is* a greenhouse effect and that too much of it can be dangerous? Nature has kindly provided, in the character of the nearest planet, a cautionary reminder. The planet Venus is a little closer to the Sun than the Earth, but its unbroken clouds are so bright that the planet actually absorbs less sunlight than the Earth. Greenhouse effect aside, its surface ought to be cooler than the Earth's. It has very closely the same size and mass as the Earth, and from all this we might naïvely conclude that it has a pleasant Earth-like environment, ultimately suitable for tourism. However, if you were to send a spacecraft through the clouds — made, by the way, largely of sulfuric acid — as the Soviet Union did in its pioneering *Venera* series of exploratory spacecraft, you would discover an extremely dense atmosphere made largely of carbon dioxide with a pressure at the surface 90 times what it is on Earth. If now you stick out a thermometer, as the *Venera* spacecraft did, you find that the temperature is some 470°C (about 900°F) — hot enough to melt tin or lead. The surface temperatures, hotter than those in the hottest household oven, are due to the greenhouse effect, largely caused by the massive carbon dioxide atmosphere. (There are also small quantities of water vapor and other infrared absorbing gases.) Venus is a practical demonstration that an increase in the abundance of greenhouse gases may have unpleasant consequences. It is a good place to point ideologically driven radio talk-show hosts who insist that the greenhouse effect is a "hoax."

As there get to be more and more humans on Earth, and as our technological powers grow still greater, we are pumping more and more infrared absorbing gases into the atmosphere. There are natural mechanisms that take these gases out of the air, but we are producing them at such a rate that we are overwhelming the removal mechanisms. Between the burning of fossil fuels and the destruction of forests (trees remove CO_2 and convert it to wood), we humans are responsible for putting about 7 billion tons of carbon dioxide into the air every year.

You can see in the figure on page 196 the increase with time of carbon dioxide in the Earth's atmosphere. The data come from the Mauna Loa atmospheric observatory in Hawaii. Hawaii is not highly industrialized and is not a place where extensive forests are being burned down (putting more CO_2 in the air). The increase in carbon dioxide with time detected over Hawaii comes from activities all over the Earth. The carbon dioxide is simply carried by the general circulation of the atmosphere worldwide — including over Hawaii. You can see that every year there's a rise and fall of carbon dioxide. That's due to deciduous trees, which, in summer, when in leaf, take CO_2 out of the atmosphere, but in winter, when leafless, do not. But superimposed on that annual oscillation is a long-term increasing trend, which is absolutely unambiguous. The CO_2 mixing ratio has now exceeded 350 parts per million — higher than it's ever been during the tenure of humans on Earth. Chlorofluorocarbon increases have been the quickest — by about 5 percent a year — because of the worldwide growth of the CFC industry, but they are now beginning to taper off.* Other greenhouse gases, methane for instance, are also building up because of our agriculture and our industry.

Well, if we know by how much greenhouse gases are building up in the atmosphere and we claim to understand what

*Again, because CFCs both deplete the ozone layer and contribute to global warming, there has been some confusion between these two very different environmental results.

the resulting infrared opacity is, shouldn't we be able to calculate the increase of temperature in recent decades as a consequence of the buildup of CO_2 and other gases? Yes, we can. But we have to be careful. We must remember that the Sun goes through an 11-year cycle, and that how much energy it puts out changes a little over its cycle. We must remember that volcanos occasionally blow their tops and inject fine sulfuric acid droplets into the stratosphere, thereby reflecting more sunlight back into space and cooling the Earth a little. A major explosion can, it is calculated, lower the world temperature by nearly a Celsius degree for a few years. We must remember that in the lower atmosphere there is a pall of tiny sulfur-containing particles from industrial smokestack pollution that — however damaging to people on the ground — also cools the Earth; as well as windblown mineral dust from disturbed soils that has a similar effect. If you make allowances for these factors and many more, if you do the best job climatologists are now capable of doing, you reach this conclusion: Over the twentieth century, due to the burning of fossil fuels, the average temperature of the Earth should have increased by a few tenths of a degree Celsius.

Naturally you would like to compare this prediction with the facts. Has the Earth's temperature increased at all, especially by this amount, during the twentieth century? Here again you must be careful. You must use temperature measurements made far from cities, because cities, through their industry, and their relative lack of vegetation, are actually hotter than the surrounding countryside. You must properly average out measurements made at different latitudes, altitudes, seasons, and times of day. You must allow for the difference between measurements on land and measurements on water. But when you do all this, the results seem consistent with the theoretical expectation.

The Earth's temperature has increased a little, less than a degree Celsius, in the twentieth century. There are substantial wiggles in the curves, noise in the global climatic signal. The ten hottest years since 1860 have all occurred in the '80s and early '90s — despite the cooling of the Earth from the 1991 explosion of the Philippine volcano Mount Pinatubo. Mount Pinatubo introduced 20 to 30 megatons of sulfur dioxide and aerosols into the Earth's atmosphere. Those materials completely circled the Earth in about three months. After only

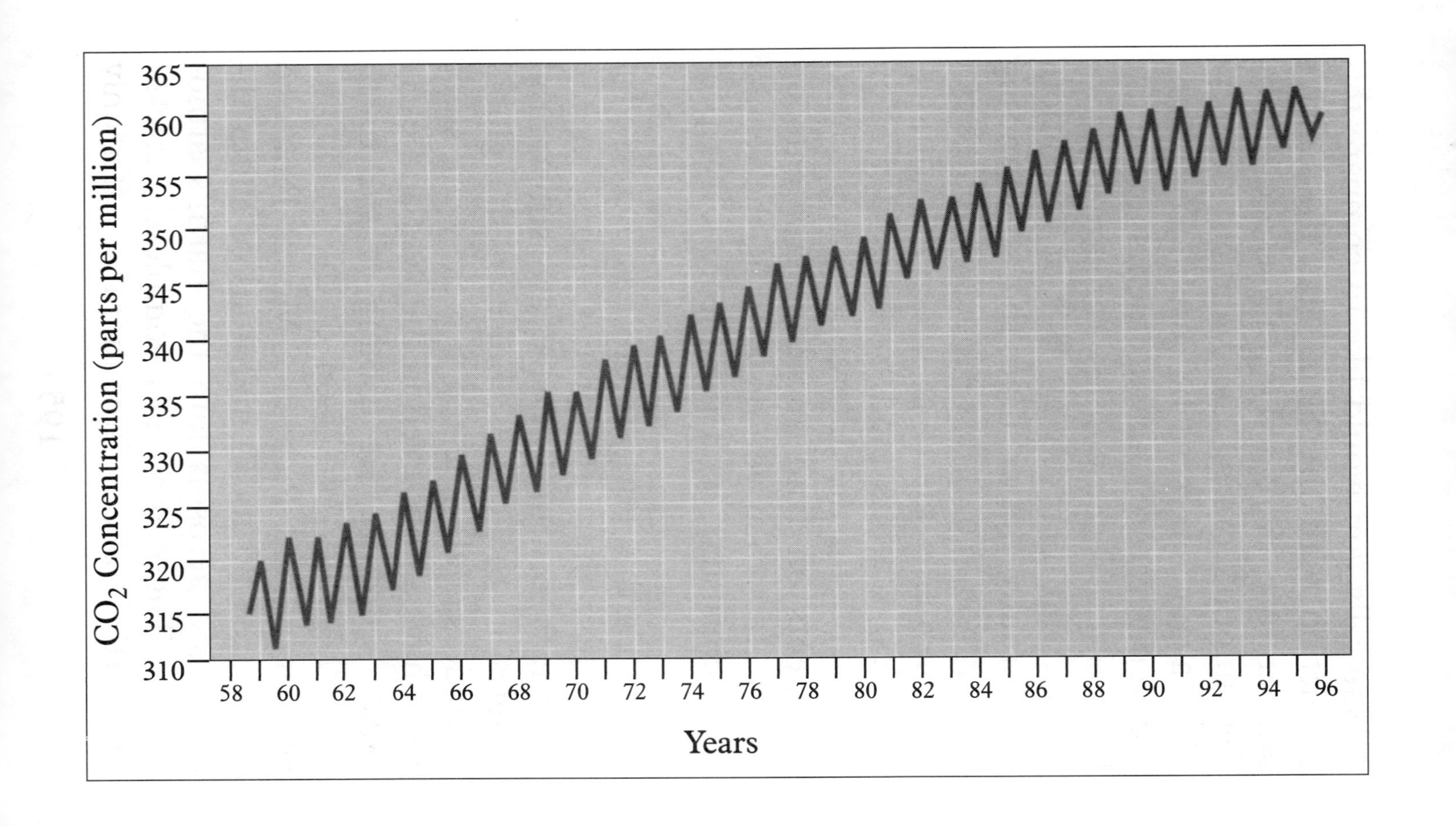
CO2 Concentration (parts per million)
365
360
355
350
345
340
335
330
325
320
315
310
58
60
62
64
66
68
70
72
74
76
78
80
82
84
86
88
90
92
94
96
Years

two months they had covered about two-fifths of the Earth's surface. This was the second most violent volcanic eruption in this century (second only to that of Mount Katmai in Alaska in 1912). If the calculations are right and there are no more big volcanic explosions in the near future, by the end of the '90s the upward trend should reassert itself. It has: 1995 was marginally the hottest year on record.

Another way to check whether the climatologists know what they're doing is to ask them to make predictions retrospectively. The Earth has gone through ice ages. There are ways of measuring how the temperature fluctuated in the past. Can they predict (or, better, postdict) the climates of the past?

Important findings on the climate history of the Earth have emerged by studying cores of ice cut and extracted from the Greenland and Antarctic ice caps. The technology for these borings comes straight from the petroleum industry; in this way, those responsible for extracting fossil fuels from the Earth have made an important contribution to clarifying the dangers of so doing. Minute physical and chemical examination of these cores reveals that the temperature of the Earth and the abundance of CO_2 in its atmosphere go up and down together — the more CO_2, the warmer the Earth. The same computer models used to understand the global temperature trends of the last few decades correctly postdict ice age climate from fluctuations in greenhouse gases in earlier times. (Of course no one is saying that there were pre-ice age civilizations that drove fuel-inefficient cars and poured enormous quantities of greenhouse gases into the atmosphere. Some variation in the amount of CO_2 happens naturally.)

In the last few hundred thousand years, the Earth has gone into and emerged out of several ice ages. Twenty thousand years ago, the city of Chicago was under a mile of ice. Today we are between ice ages, in what's called an interglacial interval. The typical temperature *difference* for the whole world between an ice age and an interglacial interval is only 3° to 6°C (equivalent to a temperature difference of 5° to 11°F). This should immediately set alarm bells ringing: A temperature change of only a few degrees can be serious business.

With this experience under their belts, this calibration of their abilities, climatologists can now try to predict just what the future climate of the Earth may be like if we keep on burning fossil fuels, if we continue to pour greenhouse gases into the atmosphere at a frenetic pace. Various scientific groups — modern equivalents of the Delphic Oracle — have employed computer models to calculate what the temperature increase ought to be, predicting how much the world temperature increases if, say, the amount of carbon dioxide in the atmosphere doubles, which it will (at the present rate of burning fossil fuels) by the end of the twenty-first century. The chief oracles are the Geophysical Fluid Dynamics Laboratory of the National Oceanic and Atmospheric Administration (NOAA) at Princeton; the Goddard Institute of Space Studies of NASA in New York; the National Center for Atmospheric Research in Boulder, Colorado; the Department of Energy's Lawrence Livermore National Laboratory in California; Oregon State University; the Hadley Center for Climate Prediction and Research in the United Kingdom; and the Max Planck Institute for Meteorology in Hamburg. They all predict that the average temperature increase will be between about 1° and 4°C. (In Fahrenheit it's about twice that.)

This is faster than any climate change observed since the rise of civilization. At the low end, developed, industrial societies, at least, might be able with a little struggle to adjust to the changed circumstances. At the high end, the climatic map of the Earth would be dramatically changed, and the consequences, both for rich and poor nations, might be catastrophic. Over much of the planet, we have confined forests and wildlife to isolated, noncontiguous areas. They will be unable to move as the climate changes. Species extinctions will be greatly accelerated. Major transplanting of crops and people will become necessary.

None of the groups claims that doubling the carbon dioxide content of the atmosphere will cool the Earth. None claims that it will heat the Earth by tens or hundreds of degrees. We have an opportunity denied to many ancient Greeks — we can go to a number of oracles and compare prophecies. When we do so, they all say more or less the same thing. The answers in fact are in good accord with the most ancient oracles on the subject — including the Swedish Nobel Prize-winning chemist Svante Arrhenius, who around the turn of the century made a similar prediction using, of course, much less sophisticated knowledge of the infrared absorption of carbon dioxide and the properties of the Earth's atmosphere. The physics used by all these groups correctly predicts the present temperature of the Earth, as well as the greenhouse effects on other planets such as Venus. Of course, there may be some simple error that everyone has missed. But surely these concordant prophecies deserve to be taken very seriously.

There are other disquieting signs. Norwegian researchers report a decrease in the extent of Arctic ice cover since 1978. Enormous rifts in the Wordie Ice Sheet in Antarctica have been evident over the same period. In January 1995, a 4,200 square kilometer piece of the Larsen Ice Shelf broke away into the Antarctic Ocean. There has been a notable retreat of mountain glaciers everywhere on Earth. Extremes of weather are increasing in many parts of the world. Sea level is continuing to rise. None of these trends by itself is compelling proof that the activities of our civilization rather than natural variability is responsible. But together, they are very worrisome.

Increasing numbers of climate experts have recently concluded that the "signature" of man-made global warming has been detected. Representatives of the 25,000 scientists of the Intergovernmental Panel on Climate Change, after an exhaustive study, concluded in 1995 that "the balance of evidence suggests there is a discernible human influence on climate." While not yet "beyond all reasonable doubt," says Michael MacCracken, director of the U.S. Global Change Research Program, the evidence "is becoming quite compelling." The observed warming "is unlikely to be caused by natural variability," says Thomas Karl of the U.S. National Climatic Data Center. "There's a 90 to 95 percent chance that we're not being fooled."

In the following sketch is a very broad perspective. At the left, it's 150,000 years ago; we have stone axes and are really pleased with ourselves for having domesticated fire. The global temperatures vary with time between deep ice ages and interglacial periods. The total amplitude of the fluctuations, from the coldest to the warmest, is about 5°C (almost 10°F). So, the curve wiggles along, and after the end of the last ice age, we have bows and arrows, domesticated animals, the origin of agriculture, sedentary life, metallic weapons, cities, police forces, taxes, exponential population growth, the Industrial Revolution, and nuclear weapons (all that last part is invented just at the extreme right of the solid curve). Then we come to the present, the end of the solid line. The dashed lines show some projections of what we're in for because of

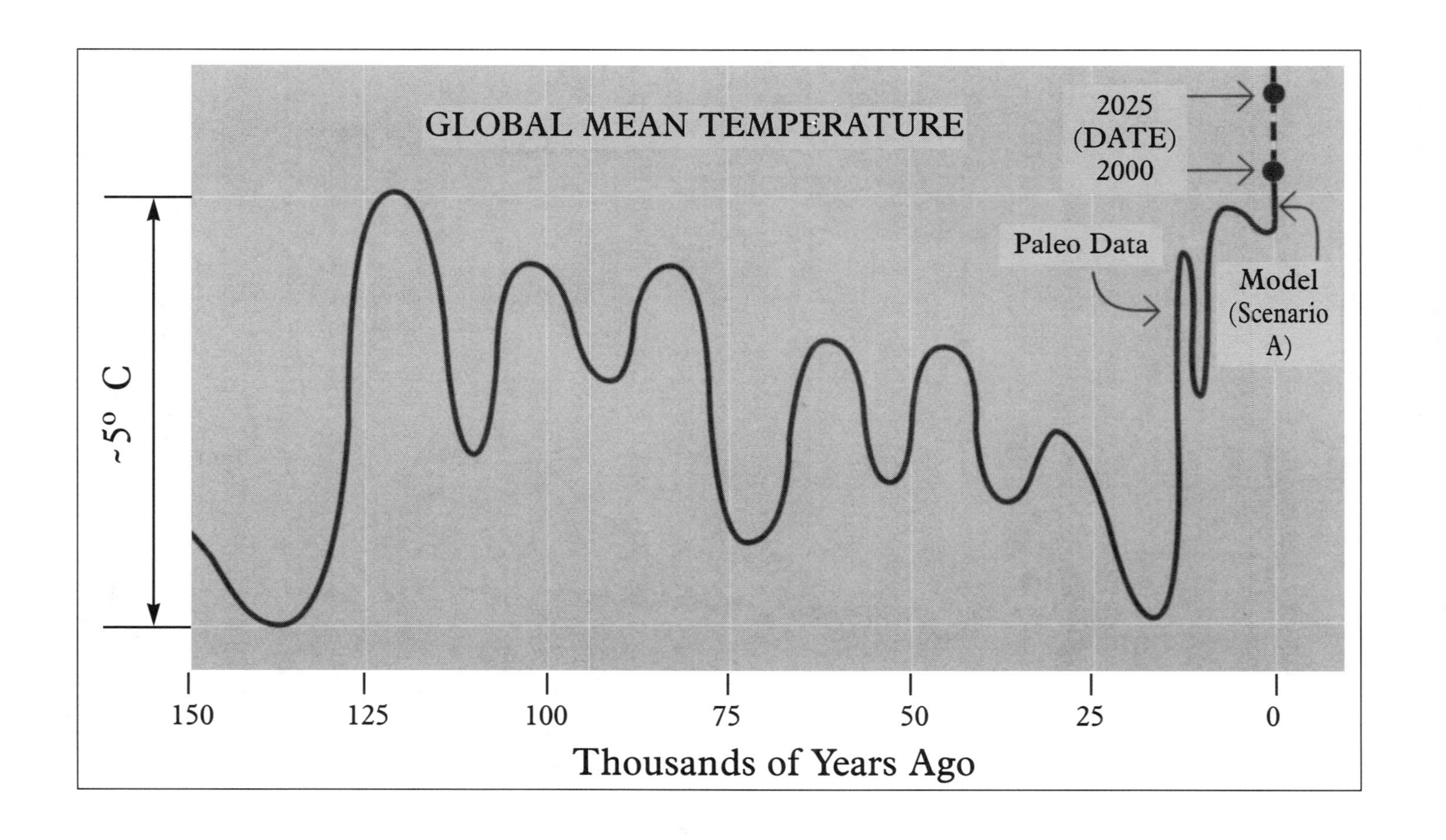
GLOBAL MEAN TEMPERATURE
2025
(DATE)
2000
Paleo Data
Model
(Scenario
A)
~5° C
150
125
100
75
50
25
0
Thousands of Years Ago

greenhouse warming. This figure makes it quite clear that the temperatures we have now (or are shortly to have if present trends continue) are not just the warmest in the last *century,* but the warmest in the last *150,000 years*. That's another measure of the magnitude of the global changes we humans are generating, and their unprecedented nature.

Global warming does not by itself make bad weather. But it does heighten the chances of having bad weather. Bad weather certainly does not require global warming, but all computer models show that global warming should be accompanied by significant increases in bad weather — severe drought inland, severe storm systems and flooding near the coasts, both much hotter and much colder weather locally, all driven by a relatively modest increment in the average planetary temperature. This is why extreme cold weather in, say, Detroit in January is not the telling refutation of global warming that some newspaper editorial pages pretend. Bad weather can be very expensive. To take a single example, the American insurance industry alone suffered a net loss of some $50 billion in the wake of a single hurricane (Andrew) in 1992, and that's only a small fraction of the total 1992 losses. Natural disasters cost the United States over $100 billion a year. The world total is much larger.

Also, changes in the weather affect animals and microbes that carry disease. Recent outbreaks of cholera, malaria, yellow fever, dengue fever, and the hantavirus pulmonary syndrome are all suspected of being related to changing weather. One recent medical estimate is that the increase in the area of the Earth occupied by the tropics and subtropics, and the resulting burgeoning population of malaria-bearing mosquitos, would by the end of the next century result in 50 to 80 million additional cases of malaria each year. Unless something is done. A 1996 U.N. scientific report argues, "If adverse population health impacts are likely to result from climate change, we do not have the usual option of seeking definitive empirical evidence before acting. A wait-and-see approach would be imprudent at best and nonsensical at worst."

The climate predicted for the next century depends on whether we put greenhouse gases into the atmosphere at the present rate, or at an accelerated rate, or at a diminished rate. The more greenhouse gases, the hotter it gets. Even assuming only moderate increases, temperatures apparently will rise significantly. But these are global averages; some places will be much colder and some much warmer. Large areas of increasing drought are predicted. In many models great food-producing areas of the world, in South and Southeast Asia, Latin America, and sub-Saharan Africa, are predicted to become hot and parched.

Some agricultural exporting nations in middle to high latitudes (the United States, Canada, Australia, for example) may actually gain at first and their exports soar. Poor nations will be most severely impacted. In the twenty-first century, in this as in many other ways, the global disparity between rich and poor may increase dramatically. Millions of people, their children starving, with very little to lose, pose a practical and serious problem for the rich — as the history of revolution teaches.

The chance of a drought-driven global agricultural crisis begins to become significant around 2050. Some scientists think that the chance of a massive worldwide agricultural failure from greenhouse warming by the year 2050 is low — perhaps only 10 percent. But of course the longer we wait, the greater the chances are. For a while some places — Canada, Siberia — might get better (if the soil is suitable for agriculture), even if the lower latitudes get worse. Wait long enough and the climate deteriorates worldwide.

As the Earth warms, sea level rises. By the end of the next century, sea level may have risen by tens of centimeters, and, just possibly, by a meter. This is partly due to the fact that seawater expands as it warms, and partly due to the melting of glacial and polar ice. As time continues, sea level rises still more. No one knows when it will happen, but eventually many populated islands in Polynesia, Melanesia, and the Indian Ocean will, according to the projections, be wholly submerged, and disappear from the face of the Earth. Understandably enough, an Alliance of Small Island States has constituted itself, militantly opposed to further increases in greenhouse gases. Devastating impacts are also predicted for Venice, Bangkok, Alexandria, New Orleans, Miami, New York City, and more generally for the highly populated areas of the Mississippi, Yangtze, Yellow, Rhine, Rhône, Po, Nile, Indus, Ganges, Niger, and Mekong rivers. Rising sea level will displace tens of millions of people in Bangladesh alone. There will be a vast new problem of environmental refugees — as populations grow, environments deteriorate, and social systems become increasingly incompetent to deal with rapid change. Where are they supposed to go? Similar problems can be anticipated for China. If we keep on with business as usual, the Earth will be warmed more every year; drought and floods both will be endemic; many more cities, provinces, and whole nations will be submerged beneath the waves — unless heroic worldwide engineering countermeasures are taken. In the longer run, still more dire consequences may follow, including the collapse of the West Antarctic ice sheet, its surge into the sea, a major global rise in sea level, and the inundation of almost all coastal cities on the planet.

Models of global warming show different effects — on temperature, drought, weather, and rising sea level, for example — becoming noticeable on different timescales, from decades to a century or two. These consequences seem so unpleasant and so expensive to fix that naturally there has been a serious effort to find something wrong with the story. Some of the efforts are motivated by nothing more than the standard scientific skepticism about all new ideas; others are motivated by the profit motive in the affected industries. One key issue is feedback.

There are both positive and negative feedbacks possible in the global climate system. Positive feedbacks are the dangerous kind. Here's an example of a positive feedback: The temperature increases a little bit because of the greenhouse effect and so some polar ice melts. But polar ice is bright compared to the open sea. As a result of that melting, then, the Earth is now very slightly darker; and because the Earth is darker, it now absorbs slightly more sunlight, so it heats some more, so it melts some more polar ice, and the process continues — perhaps to run away. That's a positive feedback. Another positive feedback: A little more CO_2 in the air heats the surface of the Earth, including the oceans, a little bit. The now warmer oceans vaporize a little more water vapor into the atmosphere. Water vapor is also a greenhouse gas, so it holds in more heat and the temperature gets higher.

Then there are negative feedbacks. They're homeostatic. An example: Heat up the Earth a little bit by putting more carbon dioxide, say, into the atmosphere. As before, this injects more water vapor into the atmosphere, but this generates more clouds. Clouds are bright; they reflect more sunlight into space, and therefore less sunlight is available to heat the Earth. The increase in temperature eventually works a decrease in temperature. Or, another possibility: Put a little more carbon dioxide into the atmosphere. Plants generally like more carbon dioxide, so they grow faster, and in growing faster, they take more carbon dioxide from the air — which in turn reduces the greenhouse effect. Negative feedbacks are like thermostats in the global climate. If, by luck, they were to be very powerful, maybe greenhouse warming would be self-limiting, and we could have the luxury of emulating Cassandra's listeners without sharing their fate.

The question is: Balance all the positive and all the negative feedbacks and where do you wind up? The answer is: Nobody is absolutely sure. Retrospective attempts to calculate global warming and cooling during the ice ages as the amount of greenhouse gases increased and decreased give the right answer. Put another way, calibrating the computer models by forcing agreement with the historical data will automatically account for all feedback mechanisms, known and unknown, in the natural climate machine. But it might be that as the Earth is pushed into climatic regimes unknown in the last 200,000 years, new feedbacks might occur of which we are ignorant. For example, much methane is sequestered in bogs (which sometimes produces the eerily beautiful dancing lights called "will-o-the-wisps"). It might begin to bubble up at an increasing pace as the Earth warms. The additional methane warms the Earth still further, and so on, another positive feedback.

Wallace Broecker of Columbia University points to the very quick warming that happened about 10,000 B.C., just before the invention of agriculture. It's so steep that, he believes, it implies an instability in the coupled ocean-atmosphere system; and that if you push the Earth's climate too hard in one direction or another, you cross a threshold, there's a kind of "bang," and the whole system runs away by itself to another stable state. He proposes that we may be teetering on just such an instability right now. This consideration only makes things worse, maybe much worse.

In any case, it's pretty clear that the faster the climate is changing, the more difficult it is for whatever homeostatic systems there are to catch up and stabilize. I wonder if we're not more likely to miss unpleasant feedbacks than comforting ones. We're not smart enough to predict everything. That's certainly clear. I think it's unlikely that the sum of what we're too ignorant to figure out will save us. Maybe it will. But would we want to bet our lives on it?

The vigor and importance of environmental issues is reflected in the meetings of professional scientific societies. For example, the American Geophysical Union is the largest organization of Earth scientists in the world. At a recent (1993) annual meeting, there was a session on previous warm episodes in Earth history, with an eye toward understanding what the consequences for global warming might be. The very first paper warns that "because future warming trends will be very rapid, there are no exact analogues for a 21st century greenhouse warming." There were four half-day sessions devoted to ozone depletion, and three sessions on the cloud/climate feedback. An additional three sessions were devoted to more general studies of the climates of the past. J. D. Mahlman of NOAA began his lecture by noting, "The discovery of the remarkably large Antarctic ozone losses in the 1980s was an event totally unpredicted by anyone." A paper from the Byrd Polar Research Center at Ohio State University offers evidence from ice cores in the West China and Peruvian glaciers for recent warming of the Earth compared to the temperatures over the last 500 years.

Considering how contentious the scientific community is, it is notable that not a single paper is offered claiming that depletion of the ozone layer or global warming are snares and delusions, or that there always was a hole in the ozone layer over Antarctica, or that global warming will be considerably less than the estimated 1° to 4°C for a doubling in the carbon dioxide abundance. The rewards for finding that there is no ozone depletion or that global warming is insignificant are very high. There are many powerful and wealthy industries and individuals that would benefit if only such contentions were true. But as the programs of scientific meetings indicate, this is probably a forlorn hope.

Our technical civilization now poses a real danger to itself. All over the world fossil fuels are simultaneously degrading respiratory health, the life of forests, lakes, coastlines, and oceans, and the world climate. Nobody intended to do any harm, surely. The captains of the fossil fuel industry were simply trying to make a profit for themselves and their shareholders, to provide a product everyone wanted, and to support the military and economic power of whatever nations they happened to be situated in. The facts that this was inadvertent, that intentions were benign, that most of us in the developed world have benefitted from our fossil fuel civilization, that many nations and many generations all contributed to the problem all suggest that this is no time for finger-pointing. No one nation or generation or industry got us into this mess, and no one nation or generation or industry can by itself get us out. If we are to prevent this climatic danger from working its worst, we will simply all have to work together, and for a long time. The principal obstacle is, of course, inertia, resistance to change — huge, worldwide, interlocking industrial, economic, and political establishments all beholden to fossil fuels, when fossil fuels are the problem. In the United States, as the evidence for the seriousness of global warming mounts, the political will to do something about it seems to be shriveling.

Chapter 12

ESCAPE FROM AMBUSH

> [P]lainly, nobody will be afraid who believes nothing can happen to him. . . . [F]ear is felt by those who believe something is likely to happen to them. . . . People do not believe this when they are, or think they are, in the midst of great prosperity, and are in consequence insolent, contemptuous and reckless. . . . [But if] they are to feel the anguish of uncertainty, there must be some faint expectation of escape.
>
> ARISTOTLE (384–322 B.C.),
> *Rhetoric*, 1382^b29

What must we do? Because the carbon dioxide that we put up into the atmosphere today will stay there for decades, even major efforts at technological self-control will do no good until a generation into the future — although the contributions by some other gases to global warming can be reduced more quickly. We need to distinguish between short-term mitigation and long-term solutions, although both are needed. We must, it seems, phase in as quickly as possible a new world energy economy that doesn't generate nearly so much greenhouse gases and other pollutants. But "as quickly as possible" will take decades at least to complete, and we must in the meantime lessen the damage, taking great care that the transition does as little damage as possible to the world's social and economic fabric, and that standards of living do not decline in consequence. The only question is whether we manage the crisis or it manages us.

Almost two out of three Americans call themselves environmentalists — according to a 1995 Gallup poll — and would give protecting the environment priority over economic growth. Most would acquiesce to increased taxes if earmarked for environmental protection. Still, it might turn out that all this is impossible — that the vested industrial interests are so powerful and consumer resistance so weak that no significant change from business-as-usual will occur until it's too late, or that the transition to a non-fossil-fuel civilization will so stress an already fragile world economy as to cause economic chaos. Plainly, we must pick our way warily. There's a natural tendency to temporize: This is unknown territory. Shouldn't we go slowly? But then we take a look at the maps of projected climate change and we recognize that we cannot temporize, that it's foolhardy to go too slowly.

The biggest CO_2 emitter on the planet is the United States. The next biggest CO_2 emitter is Russia and the other republics of the former Soviet Union. The third biggest, if we combine them, is all the developing countries together. That's a very important fact: This isn't just a problem for the highly technological nations — through slash-and-burn agriculture, burning firewood, and so on, developing countries are also making a major contribution to global warming. And the developing countries have the world's largest population growth rate. Even if they don't succeed in achieving something like the standard of living of Japan, the Pacific Crescent, and the West, these nations will constitute a steadily increasing part of the problem. Next in order of complicity is Western Europe, then China, and only then Japan, one of the most fuel-efficient nations on Earth. Again, just as the cause of global

GREENHOUSE WARMING

from the burning of coal, oil, and gas may be endangering the global environment

warming is worldwide, any solution must also be worldwide.

The scale of change necessary to address this problem at its core is nearly daunting — especially for those policymakers who are mainly interested to do things that will benefit them during their terms of office. If the required action to make things better could be subsumed in 2-, 4-, or 6-year programs, politicians would be more supportive, because then the political benefits might accrue when it's time for reelection. But 20-, 40-, or 60-year programs, where the benefits accrue not only when the politicians are out of office, but when they're dead, are politically less attractive.

Certainly we must be careful not to rush off half-cocked like Croesus and discover that at huge expense we've done something unnecessary or stupid or dangerous. But even more irresponsible is to ignore an impending catastrophe and naïvely hope it will go away. Can't we find some middle ground of policy response, which is appropriate to the seriousness of the problem, but which does not ruin us in case somehow — a negative feedback *deus ex machina*, for example — we have overestimated the severity of the matter?

Say you're designing a bridge or skyscraper. It's customary to build in, to demand, a tolerance to catastrophic failure far beyond what the likely stresses will be. Why? Because the consequences of the collapse of the bridge or skyscraper are so serious, you must be sure. You need very reliable guarantees. The same approach, I think, must be adopted for local, regional, and global environmental problems. And here, as I've said, there is great resistance, in part because large amounts of money are required from government and industry. For this reason, we will increasingly see attempts to discredit global warming. But money is also needed to truss up bridges and to reinforce skyscrapers. This is considered a normal part of the cost of building big. Designers and builders who cut corners and take no such precautions are not considered prudent capitalists because they don't waste money on implausible contingencies. They are considered criminals. There are laws to make sure bridges and skyscrapers don't fall down. Shouldn't we also have laws and moral proscriptions treating the potentially far more serious environmental issues?

I want now to offer some practical suggestions about dealing with climate change. I believe they represent the consensus of a large number of experts, although doubtless not all. They constitute only a beginning, only an attempt to mitigate the problem, but at an appropriate level of seriousness. To undo global warming and bring the Earth's climate back to what it was, say, in the 1960s will be much more difficult. The proposals are modest in another respect as well — they all have excellent reasons for being carried out, independent of the global warming issue.

With systematic monitoring of the Sun,

atmosphere, clouds, land, and ocean from space, aircraft, ships, and from the ground, using a wide range of sensing systems, we should be able to diminish the range of current uncertainty, identify feedback loops, observe regional pollution patterns and their effects, track the withering of the forests and the growth of deserts, monitor changes in the polar ice caps, in glaciers, and in the level of the oceans, examine the chemistry of the ozone layer, observe the spread of volcanic debris and its climatic consequences, and scrutinize changes in how much sunlight arrives at Earth. We have never before had such powerful tools to study and to safeguard the global environment. While spacecraft of many nations are about to play a role, the premier such tool is NASA's robotic Earth Observing System, part of its Mission to Planet Earth.

When greenhouse gases are added to the atmosphere, the Earth's climate does not respond instantaneously. Instead it seems to take about a century for two-thirds of the total effect to be felt. Thus, even if we stopped all CO_2 and other emissions tomorrow, the greenhouse effects would continue to build at least until the end of the next century. This is a powerful reason to mistrust the "wait-and-see" approach to the problem — it may be profoundly dangerous.

When there was an oil crisis in 1973–79, we raised taxes to reduce consumption, made cars smaller, and lowered the speed limits. Now that there's a glut of petroleum we've lowered taxes, made cars larger, and raised the speed limits. There's no hint of long-term thinking.

To prevent the greenhouse effect from increasing still further, the world must cut its dependence on fossil fuels by more than half. In the short term, while we're stuck with fossil fuels, we can use them much more efficiently. With 5 percent of the world's population, the United States uses nearly 25 percent of the world's energy. Automobiles are responsible for almost a third of U.S. CO_2 production. Your car emits more than its own weight in CO_2 each year. Clearly, if we can get more miles per gallon of gasoline, we'll be putting less carbon dioxide into the atmosphere. Nearly all experts agree that huge improvements in fuel efficiency are possible. Why are we — self-professed environmentalists — content with cars that get only 20 miles to the gallon? If we can drive at 40 miles per gallon, we'll be injecting only half as much CO_2 into the air; at 80 miles per gallon, only a quarter as much. This issue is typical of the emerging conflict between short-term maximizing of profits and long-term mitigation of environmental damage.

No one will buy the fuel-efficient cars, Detroit used to say; they'll have to be smaller and so more dangerous; they won't accelerate as quickly (although they certainly could go faster than the speed limits); and they'll cost more. And it is true that in the middle 1990s, Americans have been increasingly driving gas-guzzling cars and trucks at high speeds — in part because petroleum is so cheap. So the American auto industry fought and more indirectly still fights meaningful change. In 1990, for example, after great pressure from Detroit, the Senate (narrowly) rejected a bill that would have required significant improvements in fuel efficiency in American automobiles, and in 1995–96 already-mandated fuel efficiencies in a number of states were relaxed.

But downsizing cars is not required, and there are ways of making even smaller cars safer — such as new shock-absorbing structures, components that crumble or bounce, composite construction, and air bags for all seats. Apart from young men in the throes of deep testosterone intoxication, how much do we lose in forgoing the ability to exceed the speed limit in a few seconds, compared with how much we gain? There are quick-accelerating gasoline-burning cars on the road today that get 50 or more miles per gallon. The cars might cost more to buy, but certainly would cost far less to fuel: According to one U.S. Government estimate, the added expense would be recouped in only three years. As far as the claim that no one will buy such cars, this underestimates the intelligence and environmental concern of the American people — and the power of advertising let loose in support of a worthy goal.

Speed limits are established, driving licenses mandated, and many other restrictions levied on the drivers of automobiles in order to save lives. Automobiles are recognized as potentially so dangerous that it is the obligation of the government to set some limits on how they're manufactured, maintained, and driven. This is even more true once we recognize the seriousness of global warming. We've benefitted from our global civilization; can't we modify our behavior slightly to preserve it?

The design of a new, safe, fast, fuel-efficient, clean, greenhouse-responsible class of autos will spur many new technologies, and make a great deal of money for those with a technological edge. The greatest danger for the American automobile industry is that if it resists too long, the necessary new technology will be provided (and patented) by foreign competition. Detroit has a particular and parochial motivation to develop new greenhouse-responsible cars: its survival. This is not a matter of ideology or political prejudice. It follows, I believe, directly from greenhouse warming.

The three big Detroit-based auto manufacturers — prodded and partly financed by the federal government — are sluggishly but collaboratively attempting to develop a car that will achieve 80 miles a gallon, or its equivalent for cars that run off something other than gasoline. If gasoline taxes were to rise, the pressures on automakers to build more fuel-efficient cars would increase.

Lately some attitudes have been changing. General Motors has been developing an electric automobile. "You must incorporate your environmental directions into your business," advised Dennis Minano, the vice president of Corporate Affairs at GM in 1996. "Corporate America is beginning to see that it is clearly good for business. . . . There's a more sophisticated market now. People will measure you as you take environmental initiatives and incorporate them to make your business successful. They're saying, 'We won't call you green, but we'll say you have low emissions, or a good recycling program. We'll say you're environmentally responsible.' " Rhetorically, at least this is something new. But I'm waiting for that affordable 80-miles-per-gallon GM sedan.

What is an electric car? You plug it in, charge its battery, and drive away. The best such autos, made of composites, achieve a few hundred miles per charge, and have passed standard crash tests. If they are to be environmentally sound, they will have to employ something other than massive lead-acid batteries — lead is a deadly poison. And of course the charge that makes an electric car go has to come from somewhere; if, say, it's a coal-fired electric power plant, it has done nothing to mitigate global warming, whatever its contribution to reducing pollution of cities and highways.

Similar improvements can be introduced throughout the rest of the fossil-fuel economy: Coal-fired plants can be made much more efficient; large rotating industrial machinery can be designed for variable speeds; fluorescent rather than incandescent lamps can be made much more widespread. Innovations in many cases will in the long run save money and help us to extricate ourselves from a risky dependence on overseas oil. There are reasons to increase the efficiency with which we use our fuels wholly independent of our concern about global warming.

But increasing the efficiency with which we extract energy from fossil fuels isn't enough in the long run. As time goes on there will be more of us on Earth, and greater power demands. Can't we find alternatives to fossil fuels, ways of generating energy that don't produce greenhouse gases, that don't warm the Earth? One such alternative is widely known — nuclear fission, releasing not chemical energy trapped in fossil fuels, but nuclear energy locked in the heart of matter. There are no nuclear autos or airplanes, but there are nuclear ships and there certainly are nuclear power plants. The cost of electricity from nuclear power is, under ideal circumstances, about the same as that from power plants that run off coal or oil, and these plants generate no greenhouse gases. None at all. Nevertheless . . .

As Three Mile Island and Chernobyl remind us, nuclear power plants can release dangerous radioactivity, or even melt down.

They generate a witches' brew of long-lived radioactive waste that must be disposed of. "Long-lived" means *really* long-lived: The half-lives of many of the radioisotopes are centuries to millennia long. If we want to bury this stuff, we have to be sure that it will not leach out and enter into the ground-water or otherwise surprise us — and not just over a period of years, but over periods of time much longer than we have been able in the past to plan for with confidence. Otherwise, we are saying to our descendants that the wastes we will to them are *their* burden, *their* lookout, *their* danger — because we couldn't find a safer way to generate energy. (Indeed, this is just what we now do with fossil fuels.) And there's one other problem: Most nuclear power plants use or generate uranium and plutonium that can be employed to manufacture nuclear weapons. They provide a continuing temptation for rogue nations and terrorist groups.

If these issues of operational safety, radioactive waste disposal, and weapons diversion were solved, nuclear power plants might be the solution to the fossil fuel problem — or at least an important stopgap, a transitional technology until we find something better. But these conditions have not been satisfied with high confidence, and

NUCLEAR POWER

generates no greenhouse gases but presents other well-known dangers

there does not seem to be a strong prospect that they will. Continuing violations of safety standards by the nuclear power industry, systematic cover-up of those violations, and failures of enforcement by the U.S. Nuclear Regulatory Commission (driven in part by budgetary restrictions) do not inspire confidence. The burden of proof is on the nuclear power industry. Some nations such as France and Japan have made a major conversion to nuclear energy, despite these worries. Meanwhile, other nations — Sweden, for example — that had previously authorized nuclear power have now decided to phase it out.

Because of widespread public uneasiness about nuclear energy, all U.S. orders for nuclear power plants placed after 1973 have been canceled, and no new plants have been ordered since 1978. Proposals for new storage or burial sites for radioactive wastes are routinely rejected by the communities involved. The witches' brew accumulates.

There is another kind of nuclear power — not fission, where atomic nuclei are split apart, but fusion, where they are put together. In principle, fusion power plants might run off seawater — a virtually inexhaustible supply — generating no greenhouse gases, posing no dangers of radioactive waste, and wholly uninvolved with uranium and plutonium. But "in principle" doesn't count. We're in a hurry. With enormous efforts and very high technology, we are now perhaps at the point where a fusion reactor will barely generate a little more power than it uses up. The prospect for fusion power is a prospect of hypothetical, enormous, expensive, high-technology systems, which even their proponents do not imagine being available on a commercial scale for many decades. We do not have many decades. Early versions are likely to generate stupendous quantities of radioactive waste. And in any case, it's hard to imagine such systems as the answer for the developing world.

What I've talked about in the last paragraph is hot fusion — so called for a good reason: You have to bring materials up to temperatures of millions of degrees or more, as in the interior of the Sun, to make fusion go. There have also been claims for something called cold fusion, which was first announced in 1989. The apparatus sits on a desk; you put in some kinds of hydrogen, some palladium metal, run an electric current, and, it is claimed, out comes more energy than you put in, as well as neutrons and other signs of nuclear reactions. If only this were true, it might be the ideal solution to global warming. Many scientific groups all over the world have looked into cold fusion. If there's any merit to the claim, the rewards, of course, would be enormous. The overwhelming judgment of the community of physicists worldwide is that cold fusion is an illusion, a mélange of measurement errors, absence of proper control experiments, and a confusion of chemical with nuclear reactions. But there are a few groups of scientists in various nations that are continuing to look into cold fusion —

the Japanese Government, for example, has supported such research at a low level — and each such claim should be evaluated on a case-by-case basis.

Maybe some subtle, ingenious new technology — wholly unforeseen at this moment — is just around the corner that will provide tomorrow's energy. There have been surprises before. But it would be foolhardy to bet on it.

For many reasons, developing countries are particularly vulnerable to global warming. They are less able to adapt to new climates, adopt new crops, reforest, build seawalls, accommodate to drought and floods. At the same time they are especially dependent on fossil fuels. What is more natural than for China, say — with the world's second largest coal reserves — to rely on fossil fuels during its exponential industrialization? And if emissaries from Japan, Western Europe, and the United States were to go to Beijing and ask for restraint in the burning of coal and oil, wouldn't China point out that these nations did not exercise such restraint during *their* industrialization? (And anyway the 1992 Rio Framework Convention on Climate Change, ratified by 150 countries, calls for developed countries to pay the cost of limiting greenhouse gas emissions in developing countries.) Developing countries need an inexpensive, comparatively low-technology alternative to fossil fuels.

So if not fossil fuels, and not fission, and not fusion, and not some exotic new technologies, then what? In the administration of U.S. President Jimmy Carter, a solar-thermal converter was installed in the roof of the White House. Water would circulate and on sunny days in Washington, D.C., be heated by sunshine and make some contribution — perhaps 20 percent — to White House power needs, including, I suppose, Presidential showers. The more energy supplied directly by the Sun, the less energy that had to be drawn from the local electric power grid, and so the less coal and oil that needed to be spent to generate electricity for the electric power grid around the Potomac River. It didn't provide most of the energy needed, it didn't work much on cloudy days, but it was a hopeful sign of what was (and is) needed.

One of the first acts of the Presidency of Ronald Reagan was to rip the solar-thermal converter off the White House roof. It was somehow ideologically offensive. Of course it costs something to renovate the White House roof, and it costs something to buy

SOLAR ENERGY

converted to electricity is a safe, promising solution to many of the world's energy dilemmas

the additional electricity needed every day. But those responsible evidently concluded that the cost was worth the benefit. What benefit? To whom?

At the same time, Federal support for alternatives to fossil fuels and nuclear power was steeply cut, by around 90 percent. Government subsidies (including huge tax breaks) for the fossil fuel and nuclear industries remained high through the Reagan/Bush years. The Persian Gulf War of 1991 can be included, I think, in that list of subsidies. While some technical progress in alternative energy sources was made during that time — little thanks to the U.S. Government — essentially we lost 12 years. Because of how fast greenhouse gases are building up in the atmosphere, and how long their effects last, we did not have 12 years to waste. Government support for alternative energy sources is finally increasing again, but very sparingly. I'm waiting for a President to reinstall a solar-energy converter in the White House roof.

In the late 1970s there was a federal tax credit for introducing solar-thermal heaters into homes. Even in mainly cloudy places, individual homeowners who took advantage of the tax break now have abundant hot water, for which they are not charged by the utility company. The initial investment was recouped in about five years. The Reagan Administration eliminated the tax credit.

There is a range of further alternative technologies. Heat from the Earth generates electricity in Italy, Idaho, and New Zealand. Seventy-five hundred turbines, turned by wind, are generating electricity in Altamont Pass, California, with the resulting electricity sold to the Pacific Gas and Electric Company. In Traverse City, Michigan, consumers are paying somewhat higher prices for wind turbine electrical power to avoid the environmental pollution of fossil fuel electrical power plants. Many other residents are on a waiting list to sign up. With allowance for environmental costs, wind-generated electricity is now cheaper than electricity generated by coal. All of U.S. electricity use, it is estimated, could be supplied by widely spaced turbines over the windiest 10 percent of the country — largely on ranch and agricultural lands. Moreover, fuel made from green plants ("biomass conversion") might substitute for oil without increasing the greenhouse effect, because the plants take CO_2 out of the air before they're made into fuel.

But from many standpoints, it seems to me, we should be developing and supporting direct and indirect conversion of sunlight into electricity. Sunlight is inexhaustible and widely available (except in extremely cloudy places like upstate New York, where I live); has few moving parts, and needs minimal maintenance. And solar power generates neither greenhouse gases nor radioactive waste.

One solar technology is widely used: hydroelectric power plants. Water is evaporated by the heat of the Sun, rains down on highlands, courses through rivers running downhill, runs into a dam, and there turns rotating machinery that generates electricity. But there are only so many swift rivers on our planet, and in many countries what is available is inadequate to supply their energy needs.

Solar-powered cars have already competed in long-distance races. Solar power could be used for generating hydrogen fuel from water; when burned, the hydrogen simply regenerates water. There's a great deal of desert in the world that might be gainfully employed in an ecologically responsible way, for harvesting sunlight. Solar-electric or "photovoltaic" energy has been routinely used for decades to power spacecraft in the vicinity of the Earth and through the inner Solar System. Photons of light strike the cell's surface and eject electrons, whose cumulative flow is a current of electricity. These are practical, extant technologies.

But when, if ever, will solar-electric or solar-thermal technology be competitive with fossil fuels in powering homes and offices? Modern estimates, including those by the Department of Energy, are that solar technology will catch up in the decade following 2001. This is soon enough to make a real difference.

Actually, the situation is much more favorable than this. When such cost comparisons are made, the accountants keep two sets of books — one for public consumption and the other revealing the true costs. The cost of crude oil in recent years has been about $20 a barrel. But U.S. military forces have been assigned to protect foreign sources of oil, and considerable foreign aid is granted to nations largely because of oil. Why should we pretend this isn't part of the cost of oil? We abide ecologically disastrous petroleum spills (such as the Exxon *Valdez*) because of our appetite for oil. Why pretend this isn't part of the cost of oil? If we add in these additional expenses, the estimated price becomes something like $80 a barrel. If we now add the environmental costs that using this oil levies on the local and global environments, the real price might be hundreds of dollars a barrel. And when protecting the oil motivates a war, as for example the one in the Persian Gulf, the cost becomes far higher, and not just in dollars.

When anything approaching a fair accounting is attempted, it becomes clear that for many purposes solar energy (and wind, and other renewable resources) is already much cheaper than coal or oil or natural gas. The United States, and the other industrial nations, ought to be making major investments in improving the technology further and installing large arrays of solar-energy converters. But the entire Department of Energy annual budget for this technology has been about the cost of one or two of the high-performance aircraft stationed abroad to protect foreign sources of oil.

Invest now in fossil fuel efficiency or alternative energy sources, and the payoff comes years in the future. But industry and consumers and politicians, as I've mentioned, often seem focused only on the here and now. Meanwhile, pioneering American solar-energy corporations are being sold to overseas firms. Solar-electric systems are currently being demonstrated in Spain, Italy, Germany, and Japan. Even the largest commercial American solar-energy plant, in the Mojave Desert, generates only a few hundred megawatts of electricity, which it sells to Southern California Edison. Worldwide, utility planners are avoiding investments in wind turbines and solar-electric generators.

Nevertheless, there are some encouraging signs. American-made small-scale solar-electric devices are beginning to dominate the world market. (Of the three largest companies, two are controlled by Germany and Japan; the third, by U.S. fossil fuel corporations.) Tibetan herders are using solar panels to power light bulbs and radios; Somalian physicians erect solar panels on camels to keep precious vaccines cold in their trek across the desert; 50,000 small homes in India are being converted to solar-electric power. Because these systems are within the reach of the lower middle class in developing countries, and because they are nearly maintenance free, the potential market in solar rural electrification is huge.

We can and should be doing better. There should be massive federal commitment to advance this technology, and incentives offered to scientists and inventors to enter this underpopulated field. Why is "energy independence" mentioned so often as a justification for environmentally risky nuclear power plants or offshore drilling — but so rarely to justify insulation, efficient cars, or wind and solar energy? Many of these new technologies can also be used in the developing world to improve industry and standards of living without making the environmental mistakes of the developed world. If America is looking to lead the world in new basic industries, here's one on the verge of taking off.

Perhaps these alternatives can be quickly developed in a real free-market economy. Alternatively, nations might consider a small tax on fossil fuels, dedicated to developing the alternative technologies. Britain established a "Non Fossil Fuel Obligation" in 1991 amounting to 11 percent of the purchase price. In America alone, this would amount to many billions of dollars a year. But President Clinton in 1993–96 was unable to pass legislation even for a five cent per gallon gasoline tax. Perhaps future administrations can do better.

What I hope will happen is that solar-electric, wind turbine, biomass conversion, and hydrogen fuel technologies will be phased in at a respectable pace, at the same time as we greatly improve the efficiency with which we burn fossil fuels. No one is talking about abandoning fossil fuels altogether. High-intensity industrial power needs — for example, in steel foundries and aluminum smelters — are unlikely to be provided by sunlight or windmills. But if we can cut our dependence on fossil fuels by half or better, we will have done a great thing. Very different technologies are unlikely to be here soon enough to match the pace of greenhouse warming. It may well be, though, that sometime in the next century new technology will be available — cheap, clean, generating no greenhouse gases, something that can be constructed and repaired in small, poor countries around the world.

But isn't there any way to take carbon dioxide *out* of the atmosphere, to undo some of the damage we've already done? The only method of cooling down the greenhouse effect which seems both safe and reliable is to plant trees. Growing trees remove CO_2 from the air. After they're fully grown, of course, it would be missing the point to burn them; that would be undoing the very benefit we are seeking. Instead, forests should be planted and the trees, when fully grown, harvested and used, say, for building houses or furniture. Or just buried. But the amount of land worldwide that must be reforested in order for growing trees to make a major contribution is enormous, about the area of the United States. This can only be done as a collaborative enterprise of the human species. Instead, the human species is destroying an acre of forest every *second*. Everyone can plant trees — individuals, nations, industries. But especially, industry. Applied Energy Services in Arlington, Virginia, has built a coal-fired power plant in Connecticut; it is also planting trees in Guatemala that will remove from the Earth's atmosphere more carbon dioxide than the company's new facility will inject into the air over its operational lifetime. Shouldn't lumber companies plant more forests — of the fast-growing, leafy variety useful for mitigating the greenhouse effect — than they cut down? What about the coal, oil, natural gas, petroleum, and automobile industries? Shouldn't every company that puts CO_2 into the atmosphere be engaged in removing it as well? Shouldn't every citizen? What about planting trees at Christmastime? Or birthdays, weddings, and anniversaries. Our ancestors came from the trees, and we have a natural affinity for them. It is perfectly appropriate for us to plant more.

In systematically digging the corpses of ancient beings out of the Earth and burning them, we have posed a danger to ourselves. We can mitigate the danger by improving the efficiency with which we do this burning; by investing in alternative technologies (such as biomass fuels, and wind and solar energy); and by giving life to some of the same kinds of beings whose remains, ancient and modern, we are burning — the trees. These actions would provide a range of subsidiary benefits: purifying the air; slowing the extinction of species in tropical forests; reducing or eliminating oil spills; providing new technologies, new jobs, and new profits; insuring energy independence; helping the United States and other oil-dependent industrial nations to remove their uniformed sons and daughters from harm's way; and redirecting more of their military budgets to productive civilian economies.

Despite continuing resistance from the fossil fuel industries, one business has moved significantly toward taking global warming seriously — insurance companies. Violent storms and other weather extremes that are greenhouse-driven, floods, drought, and so forth might "bankrupt the industry," says the president of the Reinsurance Association of America. In May 1996, citing the fact that 6 of the 10 worst natural disasters in the history of the country occurred in the previous decade, a consortium of American insurance companies sponsored an investigation of global warming as the potential cause. German and Swiss insurance companies have lobbied for decreases in greenhouse-gas emissions. The Alliance of Small Island States has called upon the industrial nations to reduce their emission of greenhouse gases to 20 percent *below* 1990 levels by the year 2005. (Between 1990 and 1995 CO_2 emissions worldwide have increased 12 percent.) There is a new concern, at least rhetorical, in other industries about environmental responsibility — reflecting overwhelming public preference, in and to some extent beyond the developed world.

"Global warming is a grave concern likely to pose a serious threat to the very foundation of human life," said Japan, announcing that it would stabilize emissions of greenhouse gases by the year 2000. Sweden announced that it will phase out the nuclear half of its energy supply by 2010 while decreasing the CO_2 emissions of its industries by 30 percent — to be done by improving energy efficiency and by phasing in renewable energy sources; it expects to save money in the process. John Selwyn Gummer, Britain's Secretary of the Environment, declared in 1996, "We are accepting as a world community that there are to be world rules." But there is considerable resistance. The OPEC countries are opposed to reducing CO_2 emissions, because it would take a bite out of their oil revenues. Russia and many developing countries oppose it because it would be a major impediment to industrialization. The United States is the only major industrial nation taking no significant measures to counter greenhouse warming. While other nations act, it appoints committees and urges the affected industries to adopt voluntary compliance, against their short-term interest. Acting effectively on this matter, of course, will be more difficult than implementing the Montreal Protocol on CFCs and its amendments. The affected industries are much more powerful, the cost of change is much greater, and there is nothing yet as dramatic for global warming as the hole over Antarctica is in ozone depletion. Citizens will have to educate industries and governments.

CO_2 molecules, being brainless, are unable to understand the profound idea of national sovereignty. They're just blown by the wind. If they're produced in one place, they can wind up in any other place. The planet is a unit. Whatever the ideological and cultural differences, the nations of the world must work together; otherwise there will be no solution to greenhouse warming and the other global environmental problems. We are all in this greenhouse together.

Finally, in April 1993, President Bill Clinton committed the United States to do what the Bush Administration had refused to do: join about 150 other nations in signing the protocols of the Earth Summit meeting held the previous year in Rio de Janeiro. Specifically, the United States pledged that by the year 2000 it would reduce its levels of emission of carbon dioxide and other greenhouse gases to 1990 levels (1990 levels are bad enough, but at least it's a step in the right direction). Fulfilling this promise will not be easy. The United States also committed to steps to protect biological diversity in a range of ecosystems on the planet.

We cannot safely continue mindless growth in technology, and wholesale negligence about the consequences of that technology. It is well within our power to guide technology, to direct it to the benefit of everyone on Earth. Perhaps there is a kind of silver lining to these global environmental problems, because they are forcing us, willy-nilly, no matter how reluctant we may be, into a new kind of thinking — in which in some matters the well-being of the human species takes precedence over national and corporate interests. We are a resourceful species when push comes to shove. We know what to do. Out of the environmental crises of our time should come, unless we are much more foolish than I think we are, a binding up of the nations and the generations, and even the end of our long childhood.

Chapter 13

RELIGION AND SCIENCE: AN ALLIANCE

> The first day or so, we all pointed to our countries. The third or fourth day, we were pointing to our continents. By the fifth day, we were aware of only one Earth.
>
> PRINCE SULTAN BIN SALMON AL-SAUD,
> Saudi Arabian astronaut

Intelligence and tool-making were our strengths from the beginning. We used these talents to compensate for the paucity of the natural gifts — speed, flight, venom, burrowing, and the rest — freely distributed to other animals, so it seemed, and cruelly denied to us. From the time of the domestication of fire and the elaboration of stone tools, it was obvious that our skills could be used for evil as well as for good. But it was not until very recently that it dawned on us that even the benign use of our intelligence and our tools might — because we are not smart enough to foresee all consequences — put us at risk.

Now we are everywhere on Earth. We have bases in Antarctica. We visit the ocean bottoms. Twelve of us have even walked on the Moon. There are now nearly 6 billion of us, and our numbers grow by the equivalent of the population of China every decade. We have subdued the other animals and the plants (although we have been less successful with the microbes). We have domesticated many organisms and made them do our bidding. We have become, by some standards, the dominant species on Earth.

And at almost every step, we have emphasized the local over the global, the short-term over the long. We have destroyed the forests, eroded the topsoil, changed the composition of the atmosphere, depleted the protective ozone layer, tampered with the climate, poisoned the air and the waters, and made the poorest people suffer most from the deteriorating environment. We have become predators on the biosphere — full of arrogant entitlement, always taking and never giving back. And so, we are now a danger to ourselves and the other beings with whom we share the planet.

The wholesale attack on the global environment is not the fault only of profit-hungry industrialists or visionless and corrupt politicians. There is plenty of blame to share.

The tribe of scientists has played a central role. Many of us didn't even bother to think about the long-term consequences of our inventions. We have been too ready to put devastating powers into the hands of the highest bidder and the officials of whichever nation we happen to be living in. In too many cases, we have lacked a moral compass. Philosophy and science from their very beginnings have been eager, in the words of René Descartes, "to make us masters and possessors of Nature," to use science, as Francis Bacon said, to bend all of Nature into "the service of Man." Bacon talked about "Man" exercising a "right over Nature." "Nature," wrote Aristotle, "has made all animals for the sake of man." "Without man," asserted Immanuel Kant, "the whole of creation would be a mere wilderness, a thing in vain." Not so long ago we were hearing about "conquering" Nature and the "conquest" of space — as if Nature and the Cosmos were enemies to be vanquished.

The religious tribe also has played a central role. Western sects held that just as we must submit to God, so the rest of Nature must submit to us. In modern times especially, we seem more dedicated to the second half of this proposition than the first. In the real and palpable world, as revealed by what we do and not what we say, many humans seemingly aspire to be lords of Creation — with an occasional token bow, as required by social convention, to whatever gods may lately be fashionable. Descartes and Bacon were profoundly influenced by religion. The notion of "us against Nature" is a legacy of our religious traditions. In the Book of Genesis, God gives humans "dominion . . . over every living thing," and the "fear" and "dread" of us is to be upon "every beast." Man is urged to "subdue" Nature, and "subdue" is translated from a Hebrew word with strong military connotations. There is much else in the Bible — and in the medieval Christian tradition out of which modern science emerged — along similar lines. Islam, by contrast, is disinclined to declare Nature an enemy.

Of course, both science and religion are complex and multilayered structures, embracing many different, even contradictory, opinions. It is scientists who discovered and called the world's attention to the environmental crises, and there are scientists who, at considerable cost to themselves, refused to work on inventions that might harm their fellows. And it is religion that first articulated the imperative to revere living things.

True, there is nothing in the Judeo-Christian-Muslim tradition that approaches the cherishing of Nature in the Hindu-Buddhist-Jain tradition or among Native Americans. Indeed, both Western religion and Western science have gone out of their way to assert that Nature is just the setting and not the story, that viewing Nature as sacred is sacrilege.

Nevertheless, there is a clear religious counterpoint: The natural world is a creation of God, put here for purposes separate from the glorification of "Man" and deserving, therefore, of respect and care in its own right, and not just because of its utility for us. A poignant metaphor of "stewardship" has emerged, especially recently — the idea that humans are the caretakers of the Earth, put here for the purpose and accountable, now and into the indefinite future, to the Landlord.

Of course, life on Earth got along pretty well for 4 billion years without "stewards."

Trilobites and dinosaurs, who were each around for more than a hundred million years, might be amused at a species here only a thousandth as long deciding to appoint itself the guardian of life on Earth. That species is itself the danger. Human stewards are needed, these religions recognize, to protect the Earth from humans.

The methods and ethos of science and religion are profoundly different. Religion frequently asks us to believe without question, even (or especially) in the absence of hard evidence. Indeed, this is the central meaning of faith. Science asks us to take nothing on faith, to be wary of our penchant for self-deception, to reject anecdotal evidence. Science considers deep skepticism a prime virtue. Religion often sees it as a barrier to enlightenment. So, for centuries, there has been a conflict between the two fields — the discoveries of science challenging religious dogmas, and religion attempting to ignore or suppress the disquieting findings.

But times have changed. Many religions are now comfortable with an Earth that goes around the Sun, with an Earth that's 4.5 billion years old, with evolution, and with the other discoveries of modern science. Pope John Paul II has said, "Science can purify religion from error and superstition; religion can purify science from idolatry and false absolutes. Each can draw the other into a wider world, a world in which both can flourish. . . . Such bridging ministries must be nurtured and encouraged."

Nowhere is this more clear than in the current environmental crisis. No matter whose responsibility the crisis mainly is, there's no way out of it without understanding the dangers and their mechanisms, and without a deep devotion to the long-term well-being of our species and our planet — that is, pretty closely, without the central involvement of both science and religion.

It has been my good fortune to participate in an extraordinary sequence of gatherings throughout the world: The leaders of the planet's religions have met with scientists and legislators from many nations to try to deal with the rapidly worsening world environmental crisis.

Representatives of nearly 100 nations were present at the "Global Forum of Spiritual and Parliamentary Leaders" conferences at Oxford in April 1988 and in Moscow in January 1990. Standing under an immense photograph of the Earth from space, I found myself looking out over a diversely costumed representation of the wondrous variety of our species: Mother Teresa and the Cardinal Archbishop of Vienna, the Archbishop of Canterbury, the chief rabbis of Romania and the United Kingdom, the Grand Mufti of Syria, the Metropolitan of Moscow, an elder of the Onondaga Nation, the high priest of the Sacred Forest of Togo, the Dalai Lama, Jain priests resplendent in their white robes, turbaned Sikhs, Hindu swamis, Buddhist abbots, Shinto priests, evangelical Protestants, the Primate of the Armenian Church, a "Living Buddha" from China, the bishops of Stockholm and Harare, metropolitans of the Orthodox Churches, the Chief of Chiefs of the Six Nations of the Iroquois Confederacy — and, joining them, the Secretary-General of the United Nations; the Prime Minister of Norway; the founder of a Kenyan women's movement to replant the forests; the President of the World Watch Institute; the directors of the United Nations' Children's Fund, its Population Fund, and UNESCO; the Soviet Minister of the Environment; and parliamentarians from dozens of nations, including U.S. Senators and Representatives and a Vice-President-to-be. These meetings were mainly organized by one person, a former U.N. official, Akio Matsumura.

I remember the 1,300 delegates assembled in St. George's Hall in the Kremlin to hear an address by Mikhail Gorbachev. The session was opened by a venerable Vedic monk, representing one of the oldest religious traditions on Earth, inviting the multitude to chant the sacred syllable "Om." As nearly as I could tell, Foreign Minister Eduard Shevardnadze went along with the "Om," but Mikhail Gorbachev restrained himself. (An immense milky-white statue of Lenin, hand outstretched, loomed nearby.)

That same day, ten Jewish delegates, finding themselves in the Kremlin at sundown on a Friday, performed the first Jewish religious service ever held there. I remember the Grand Mufti of Syria stressing, to the surprise and delight of many, the importance in Islam of "birth control for the global welfare, without exploiting it at the expense of one nationality over another." Several speakers quoted the Native American saying, "We have not inherited the Earth from our ancestors, but have borrowed it from our children."

The interconnectedness of all human beings was a theme constantly stressed. We heard a secular parable, which asked us to imagine our species as a village of 100 families. Then, 65 families in our village are illiterate, and 90 do not speak English, 70 have no drinking water at home, 80 have no members who have ever flown in an airplane. Seven families own 60 percent of the land and consume 80 percent of all the available energy. They have all the luxuries. Sixty families are crowded onto 10 percent of the land. Only one family has any member with a university education. And the air and the water, the climate and the blistering sunlight, are all getting worse. What is our common responsibility?

At the Moscow conference, an appeal signed by a number of distinguished scientists was presented to world religious leaders. Their response was overwhelmingly positive. The meeting ended with a plan of action that included these sentences:

> This gathering is not just an event but a step in an ongoing process in which we are irrevocably involved. So now we return home pledged to act as devoted participants in this process, nothing less than emissaries for fundamental change in attitudes and practices that have pushed our world to a perilous brink.

* * *

Religious leaders in many nations have begun to move into action. Major steps have been taken by the U.S. Catholic Conference, the Episcopal Church, the United Church of Christ, evangelical Christians, leaders of the Jewish community, and many other groups. As a catalyst of this process, a Joint Appeal of Science and Religion for the Environment was established, chaired by the Very Reverend James Parks Morton, dean of the Cathedral of St. John the Divine, and myself. Vice President Al Gore, then a U.S. Senator, played a central role. At an exploratory meeting of scientists and leaders of the major American denominations, held in New York in June 1991, it became clear that there was a great deal of common ground:

> Much would tempt us to deny or push aside this global environmental crisis and refuse even to consider the fundamental changes of human behavior required to address it. But we religious leaders accept a prophetic responsibility to make known the full dimensions of this challenge, and what is required to address it, to the many millions we reach, teach and counsel.

We intend to be informed participants in discussions of these issues and to contribute our views on the moral and ethical imperative for developing national and international policy responses. But we declare here and now that steps must be taken toward: accelerated phaseout of ozone-depleting chemicals; much more efficient use of fossil fuels and the development of a non-fossil fuel economy; preservation of tropical forests and other measures to protect continued biological diversity; and concerted efforts to slow the dramatic and dangerous growth in world population through empowering both women and men, encouraging economic self-sufficiency, and making family education programs available to all who may consider them on a strictly voluntary basis.

We believe a consensus now exists, at the highest level of leadership across a significant spectrum of religious traditions, that the cause of environmental integrity and justice must occupy a position of utmost priority for people of faith. Response to this issue can and must cross traditional religious and political lines. It has the potential to unify and renew religious life.

The last phrase of the middle paragraph represents a tortuous compromise with the Roman Catholic delegation, opposed not only to describing birth control methods, but even to uttering the words "birth control."

By 1993, the Joint Appeal had evolved into The National Religious Partnership for the Environment, a coalition of the Catholic, Jewish, mainline Protestant, Eastern Orthodox, historic black church, and evangelical Christian communities. Using material prepared by the Partnership's Science Office, the participating groups — both individually and collectively — have begun to exert considerable influence. Many religious communities previously without national environmental programs or offices are now described as "fully committed to this enterprise." Manuals on environmental education and action have reached over 100,000 religious congregations, representing tens of millions of Americans. Thousands of clergy and lay leaders have participated in regional training, and thousands of congregational environmental initiatives have been documented. State and national legislators have been lobbied, the media briefed, seminarians apprised, sermons delivered. As a more or less random example, in January 1996, the Evangelical Environmental Network — the Partnership's constituent organization from the evangelical Christian community — lobbied Congress in support of the Endangered Species Act (which is itself endangered). The grounds? A spokesman explained that while the evangelicals were "not scientists," they could "make the case" on theological grounds: Laws protecting endangered species were described as "the Noah's Ark of our day." The Partnership's fundamental tenet, "that environmental protection must now be a central component of faith life," is apparently being widely accepted. There is one major initiative that the Partnership has not yet broached: an outreach to parishioners who are executives of major industries that affect the environment. I very much hope that this will be attempted.

The present world environmental crisis is not yet a disaster. Not yet. As in other crises, it has a potential to draw forth previously untapped and even unimagined powers of cooperation, ingenuity, and commitment. Science and religion may differ about how the Earth was made, but we can agree that protecting it merits our profound attention and loving care.

THE APPEAL

What follows is the January 1990 text, sent by scientists to religious leaders, of "Preserving and Cherishing the Earth: An Appeal for Joint Commitment in Science and Religion."

The Earth is the birthplace of our species and, so far as we know, our only home. When our numbers were small and our technology feeble, we were powerless to influence the environment of our world. But today, suddenly, almost without anyone noticing, our numbers have become immense and our technology has achieved vast, even awesome, powers. Intentionally or inadvertently, we are now able to make devastating changes in the global environment — an environment to which we and all the other beings with which we share the Earth are meticulously and exquisitely adapted.

We are now threatened by self-inflicted, swiftly moving environmental alterations about whose long-term biological and ecological consequences we are still painfully ignorant — depletion of the protective ozone layer; a global warming unprecedented in the last 150 millennia; the obliteration of an acre of forest every second; the rapid-fire extinction of species; and the prospect of a global nuclear war that would put at risk most of the population of the Earth. There may well be other such dangers of which, in our ignorance, we are still unaware. Individually and cumulatively they represent a trap being set for the human species, a trap we are setting for ourselves. However principled and lofty (or naïve and shortsighted) the justifications may have been for the activities that brought forth these dangers, separately and together they now imperil our species and many others. We are close to committing — many would argue we are already committing — what in religious language is sometimes called Crimes against Creation.

By their very nature these assaults on the environment were not caused by any one political group or any one generation. Intrinsically, they are transnational, transgenerational, and transideological. So are all conceivable solutions. To escape these traps requires a perspective that embraces the peoples of the planet and all the generations yet to come.

Problems of such magnitude, and solutions demanding so broad a perspective, must be recognized from the outset as having a religious as well as a scientific dimension. Mindful of our common responsibility, we scientists — many of us long engaged in combating the environmental crisis — urgently appeal to the world religious community to commit, in word and deed, and as boldly as is required, to preserve the environment of the Earth.

Some of the short-term mitigations of these dangers — such as greater energy efficiency, rapid banning of chlorofluorocarbons, or modest reductions in the nuclear arsenals — are comparatively easy and at some level are already under way. But other, more far-reaching, more long-term, more effective approaches will encounter widespread inertia, denial, and resistance. In this category are conversion from fossil fuels to a nonpolluting energy economy, a continuing swift reversal of the nuclear arms race, and a voluntary halt to world population growth — without which many of the other approaches to preserving the environment will be nullified.

As on issues of peace, human rights, and social justice, religious institutions can here too be a strong force encouraging national and international initiatives in both the private and public sectors, and in the diverse worlds of commerce, education, culture, and mass communication.

The environmental crisis requires radical changes not only in public policy, but also in individual behavior. The historical record makes it clear that religious teaching, example, and leadership are powerfully able to influence personal conduct and commitment.

As scientists, many of us have had profound experiences of awe and reverence before the Universe. We understand that what is regarded as sacred is more likely to be treated with care and respect. Efforts to safeguard and cherish the environment need to be infused with a vision of the sacred. At the same time, a much wider and deeper understanding of science and technology is needed. If we do not understand the problem, it is unlikely we will be able to fix it. Thus there is a vital role for both religion and science.

We know that the well-being of our

> planetary environment is already a source of profound concern in your councils and congregations. We hope this Appeal will encourage a spirit of common cause and joint action to help preserve the Earth.

The response to this Scientists' Appeal on the Environment was soon after signed by hundreds of spiritual leaders from 83 countries, including 37 heads of national and international religious bodies. Among them are the general secretaries of the World Muslim League and World Council of Churches, the vice president of the World Jewish Congress, the Catholicos of All Armenians, Metropolitan Pitirim of Russia, the grand muftis of Syria and the former Yugoslavia, the presiding bishops of all the Christian churches of China and of the Episcopal, Lutheran, Methodist, and Mennonite churches in the United States, as well as 50 cardinals, lamas, archbishops, head rabbis, patriarchs, mullahs, and bishops of major world cities. They said:

> We are moved by the Appeal's spirit and challenged by its substance. We share its sense of urgency. This invitation to collaboration marks a unique

moment and opportunity in the relationship of science and religion.

Many in the religious community have followed with growing alarm reports of threats to the well-being of our planet's environment such as those set forth in the Appeal. The scientific community has done humankind a great service by bringing forth evidence of these perils. We encourage continued scrupulous investigation and must take account of its results in all our deliberations and declarations regarding the human condition.

We believe the environmental crisis is intrinsically religious. All faith traditions and teachings firmly instruct us to revere and care for the natural world. Yet sacred creation is being violated and is in ultimate jeopardy as a result of long-standing human behavior. A religious response is essential to reverse such long-standing patterns of neglect and exploitation.

For these reasons, we welcome the Scientists' Appeal and are eager to explore as soon as possible concrete, specific forms of collaboration and action. The Earth itself calls us to new levels of joint commitment.

Part III

WHERE HEARTS AND MINDS COLLIDE

Chapter 14

THE COMMON ENEMY

I am not a pessimist. To perceive evil where it exists is, in my opinion, a form of optimism.

ROBERTO ROSSELLINI

It is only in the moment of time represented by the present century that one species has acquired the power to alter the nature of the world.

RACHEL CARSON,
Silent Spring (1962)

INTRODUCTION

In 1988 a unique opportunity was presented to me. I was invited to write an article on the relationship between the United States and the then Soviet Union that would be published, more or less simultaneously, in the most widely circulated publications of both countries. This was at a time when Mikhail Gorbachev was feeling his way on giving Soviet citizens the right to express their opinions freely. Some recall it as a time when the administration of Ronald Reagan was slowly modifying its pointed Cold War posture. I thought such an article might be able to do a little good. What's more, at a recent "summit" meeting, Mr. Reagan had commented that if only there were a peril of alien invasion of the Earth, it would be much easier for the United States and the Soviet Union to work together. This seemed to give my piece an organizing principle. I intended the article to be provocative to citizens of both countries and required assurances from both sides that there would be no censorship. Both the editor of *Parade*, Walter Anderson, and the editor of *Ogonyok*, Vitaly Korotich, readily agreed. Called "The Common Enemy," the article duly appeared in the February 7, 1988 issue of *Parade*, and in the March 12-19, 1988, issue of *Ogonyok*. It was subsequently reprinted in *The Congressional Record*, won the Olive Branch Award of New York University in 1989, and was widely discussed in both countries.

The controversial matters in the article were handled straightforwardly by *Parade*, with the following introduction:

The following article, which also is scheduled to appear in its entirety in Ogonyok, *the most popular magazine in the Soviet Union, explores the relationship between our two nations. Citizens of both countries may find some of Carl Sagan's insights uncomfortable and even provocative because, fundamentally, he challenges popular views of each nation's history. The editors of Parade hope that this analysis, as it is read here and in the Soviet Union, constitutes a first step to achieve the very goals the author describes.*

But things were not nearly so simple even in the liberalizing Soviet Union of 1988. Korotich had bought a pig in a poke, and when he saw my critical comments on Soviet history and policy, he felt obliged to seek the guidance of higher authority. The responsibility for the article's content, as it appeared in *Ogonyok*, seems to have rested ultimately with Dr. Georgi Arbatov — Director of the Institute of the USA and Canada of the then Soviet Academy of Sciences, a member of the Central Committee of the Communist Party, and a close advisor to Gorbachev. Arbatov and I had privately held several political conversations that had surprised me in their frankness and candor. While in a way it is gratifying to see how much of the text was left untouched, it is also instructive to note what changes were made, what thoughts were considered too dangerous for the average Soviet citizen. So at the end of the article I've indicated the most interesting changes. They certainly do amount to censorship.

THE ARTICLE

If only, said the American President to the Soviet General Secretary, extraterrestrials were about to invade — then our two countries could unite against the common enemy. Indeed, there are many instances when deadly adversaries, at one another's throats for generations, put their differences aside to confront a still more urgent threat: the Greek city-states against the Persians; the Russians and the Polovtsys (who once had sacked Kiev) against the Mongols; or, for that matter, the Americans and the Soviets against the Nazis.

An alien invasion is, of course, unlikely. But there *is* a common enemy — in fact, a range of common enemies, some of unprecedented menace, each unique to our time. They derive from our growing technological powers and from our reluctance to forgo perceived short-term advantages for the longer-term well-being of our species.

The innocent act of burning coal and other fossil fuels increases the carbon dioxide greenhouse effect and raises the temperature of the Earth, so that in less than a century, according to some projections, the American Midwest and the Soviet Ukraine — current breadbaskets of the world — may be converted into something approaching scrub deserts. Inert, apparently harmless gases used in refrigeration deplete the protective ozone layer; they increase the amount of deadly ultraviolet radiation from the Sun that reaches the surface of the Earth, destroying vast numbers of unprotected microorganisms that lie at the base of a poorly understood food chain — at the top of which precariously teeter we. American industrial pollution destroys forests in Canada. A Soviet nuclear reactor accident endangers the ancient culture of Lapland. Raging epidemic disease spreads worldwide, accelerated by modern transportation technology. And inevitably there will be other perils that with our usual bumbling, short-term focus, we have not yet even discovered.

The nuclear arms race, jointly pioneered by the United States and the Soviet Union, has now booby-trapped the planet with some 60,000 nuclear weapons — far more than enough to obliterate both nations, to jeopardize the global civilization, and perhaps even to end the million-year-long human experiment. Despite indignant protestations of peaceable intent and solemn treaty obligations to reverse the nuclear arms race, the United States and the Soviet Union together still somehow manage to build enough new nuclear weapons each year to destroy every sizable city on the planet. When asked for justification, each earnestly points to the other. In the wake of the *Challenger* space shuttle and Chernobyl nuclear power plant disasters, we are reminded that catastrophic failures in high technology can occur despite our best efforts. In the century of Hitler, we recognize that madmen can achieve absolute control over modern industrial states. It is only a matter of time until there occurs some unanticipated subtle error in the machinery of mass destruction, or some crucial communications failure, or some emotional crisis in an already burdened national leader. Overall, the human species spends almost $1 trillion a year, most of it by the United States and the Soviet Union, in preparations for intimidation and war. Perhaps, in retrospect, there would be little motivation even for malevolent extraterrestrials to attack the

Earth; perhaps, after a preliminary survey, they might decide it is more expedient just to be patient for a little while and wait for us to self-destruct.

We are at risk. We do not need alien invaders. We have all by ourselves generated sufficient dangers. But they are unseen dangers, seemingly far removed from everyday life, requiring careful thought to understand, and involving transparent gases, invisible radiation, nuclear weapons that almost no one has actually witnessed in use — not a foreign army intent on plunder, slavery, rape, and murder. Our common enemies are harder to personify, more difficult to hate than a Shahanshah, a Khan, or a Führer. And joining forces against these new enemies requires us to make courageous efforts at self-knowledge, because we ourselves — all the nations of the Earth, but especially the United States and the Soviet Union — bear responsibility for the perils we now face.

Our two nations are tapestries woven from a rich diversity of ethnic and cultural threads. Militarily, we are the most powerful nations on Earth. We are advocates of the proposition that science and technology can make a better life for all. We share a stated belief in the right of the people to rule themselves. Our systems of government were born in historic revolutions against injustice, despotism, incompetence, and superstition. We come from revolutionaries who accomplished the impossible — freeing us from tyrannies entrenched for centuries and thought to be divinely ordained. What will it take to free us from the trap we have set for ourselves?

Each side has a long list of deeply resented abuses committed by the other — some imaginary, most, in varying degrees, real. Every time there is an abuse by one side, you can be sure of some compensatory abuse by the other. Both nations are full of wounded pride and professed moral rectitude. Each knows in excruciating detail the most minor malefactions of the other but hardly even glimpses its own sins and the suffering its own policies have caused. On each side, of course, there are good and honest people who see the dangers their national policies have created — people who long, as a matter of elementary decency and simple survival, to put things right. But there are also, on both sides, people gripped by a hatred and fear intentionally fanned by the respective agencies of national propaganda, people who believe their adversaries are beyond redemption, people who seek confrontation. The hard-liners on each side encourage one another. They owe their credibility and their power to one another. They need one another. They are locked in a deadly embrace.

If no one else, alien or human, can extricate us from this deadly embrace, then there is only one remaining alternative: However painful it may be, we will just have to do it ourselves. A good start is to examine the historical facts as they might be viewed by the other side — or by posterity, if any. Imagine first a Soviet observer considering some of the events of American history: The United States, founded on principles of freedom and liberty, was the last major nation to end chattel slavery; many of its founding fathers — George Washington and Thomas Jefferson among them — were slave owners; and racism was legally protected for a century after the slaves were freed. The United States has systematically violated more than 300 treaties it signed guaranteeing some of the rights of the original inhabitants of the country. In 1899, two years before becoming President, Theodore Roosevelt, in a widely admired speech, advocated "righteous war" as the sole means of achieving "national greatness." The United States invaded the Soviet Union in

1918 in an unsuccessful attempt to undo the Bolshevik Revolution. The United States invented nuclear weapons and was the first and only nation to explode them against civilian populations — killing hundreds of thousands of men, women, and children in the process. The United States had operational plans for the nuclear annihilation of the Soviet Union before there even was a Soviet nuclear weapon, and it has been the chief innovator in the continuing nuclear arms race. The many recent contradictions between theory and practice in the United States include the present [Reagan] Administration, in high moral dudgeon, warning its allies not to sell arms to terrorist Iran while secretly doing just that; conducting worldwide covert wars in the name of democracy while opposing effective economic sanctions against a South African regime in which the vast majority of citizens have no political rights at all; being outraged at Iranian mining of the Persian Gulf as a violation of international law, while it has itself mined Nicaraguan harbors and subsequently fled from the jurisdiction of the World Court; vilifying Libya for killing children and in retaliation killing children; and denouncing the treatment of minorities in the Soviet Union, while America has more young black men in jail than in college. This is not just a matter of mean-spirited Soviet propaganda. Even people congenially disposed toward the United States may feel grave reservations about its real intentions, especially when Americans are reluctant to acknowledge the uncomfortable facts of their history.

Now imagine a Western observer considering some of the events in Soviet history. Marshal Tukhachevsky's marching orders on July 2, 1920, were, "On our bayonets we will bring peace and happiness to toiling humanity. Forward to the West!" Shortly after, V. I. Lenin, in conversation with French delegates, remarked: "Yes, Soviet troops are in Warsaw. Soon Germany will be ours. We will reconquer Hungary. The Balkans will rise against capitalism. Italy will tremble. Bourgeois Europe is cracking at all its seams in this storm." Then contemplate the millions of Soviet citizens killed by Stalin's deliberate policy in the years between 1929 and World War II — in forced collectivization, mass deportation of peasants, the resulting famine of 1932–33, and the great purges (in which almost the entire Communist Party hierarchy over the age of 35 was arrested and executed, and during which a new constitution that allegedly safeguarded the rights of Soviet citizens was proudly proclaimed). Then consider Stalin's decapitation of the Red Army, the secret protocol to his nonaggression pact with Hitler, and his refusal to believe in a Nazi invasion of the U.S.S.R. even after it had begun — and how many millions more were killed in consequence. Think of Soviet restrictions on civil liberties, freedom of expression, and the right to emigrate, and continuing endemic anti-Semitism and religious persecution. If, then, shortly after your nation is established, your highest military and civilian leaders boast about their intentions of invading neighboring states; if your absolute leader for almost half your history is someone who methodically killed millions of his own people; if, even now, your coins display your national symbol emblazoned over the whole world — you can understand that citizens of other nations, even those with peaceful or credulous dispositions, might be skeptical of your present good intentions, however sincere and genuine they may be. This is not merely a matter of mean-spirited American propaganda. The problem is compounded if you pretend such things never happened.

"No nation can be free if it oppresses other nations," wrote Friedrich Engels. At the London conference of 1903, Lenin advocated the "complete right of self-determination of all nations." The same principles were uttered in almost exactly the same language by Woodrow Wilson and by many other American statesmen. But for both nations the facts speak otherwise. The Soviet Union has forcibly annexed Latvia, Lithuania, Estonia, and parts of Finland, Poland, and Romania; occupied and brought under Communist control Poland, Romania, Hungary, Mongolia, Bulgaria, Czechoslovakia, East Germany, and Afghanistan; and suppressed the East German workers' uprising of 1953, the Hungarian Revolution of 1956, and the Czech attempt to introduce *glasnost* and *perestroika* in 1968. Excluding World Wars and expeditions to suppress piracy or the slave trade, the United States has made armed invasions and interventions in other countries on more than 130 separate occasions,* including China (on 18 separate occasions), Mexico (13), Nicaragua and Panama (9 each), Honduras (7), Colombia and Turkey (6 each), the Dominican Republic, Korea, and Japan (5

*This list, which occasioned some surprise when published in America, is based on compilations by the House Armed Services Committee.

each), Argentina, Cuba, Haiti, the Kingdom of Hawaii, and Samoa (4 each), Uruguay and Fiji (3 each), Guatemala, Lebanon, the Soviet Union, and Sumatra (2 each), Grenada, Puerto Rico, Brazil, Chile, Morocco, Egypt, Ivory Coast, Syria, Iraq, Peru, Formosa, the Philippines, Cambodia, Laos, and Vietnam. Most of these incursions were small-scale efforts to maintain compliant governments or to protect American property and business interests; but some were much larger, more prolonged, and on far deadlier scales.

United States armed forces were intervening in Latin America not only before the Bolshevik Revolution but also before the *Communist Manifesto* — which makes the anti-Communist justification for American intervention in Nicaragua a little difficult to rationalize; the deficiencies of the argument would be better understood, however, had the Soviet Union not been in the habit of gobbling up other countries. The American invasion of Southeast Asia — of nations that never had harmed or threatened the United States — killed 58,000 Americans and more than a million Asians; the U.S. dropped 7.5 megatons of high explosives and produced an ecological and economic chaos from which the region still has not recovered.

More than 100,000 Soviet troops have, since 1979, been occupying Afghanistan — a nation with a lower per capita income than Haiti — with atrocities still largely untold (because Soviets are much more successful than Americans in excluding independent reporters from their war zones).

Habitual enmity is corrupting and self-sustaining. If it falters, it can easily be revived by reminding us of past abuses, by contriving an atrocity or a military incident, by announcing that the adversary has deployed some dangerous new weapon, or merely by taunts of naïveté or disloyalty when domestic political opinion becomes uncomfortably evenhanded. For many Americans, communism means poverty, backwardness, the Gulag for speaking one's mind, a ruthless crushing of the human spirit, and a thirst to conquer the world. For many Soviets, capitalism means heartless and insatiable greed, racism, war, economic instability, and a worldwide conspiracy of the rich against the poor. These are caricatures — but not wholly so — and over the years Soviet and American actions have given them some credence and plausibility.

These caricatures persist because they are

partly true, but also because they are useful. If there is an implacable enemy, then bureaucrats have a ready excuse for why prices go up, why consumer goods are unavailable, why the nation is noncompetitive in world markets, why there are large numbers of unemployed and homeless people, or why criticism of leaders is unpatriotic and impermissible — and especially why so supreme an evil as nuclear weapons must be deployed in the tens of thousands. But if the adversary is insufficiently wicked, the incompetence and failed vision of government officials cannot be so easily ignored. Bureaucrats have motives for inventing enemies and exaggerating their misdeeds.

Each nation has military and intelligence establishments that evaluate the danger posed by the other side. These establishments have a vested interest in large military and intelligence expenditures. Thus, they must grapple with a continuing crisis of conscience — the clear incentive to exaggerate the adversary's capabilities and intentions. When they succumb, they call it necessary prudence; but whatever they call it, it propels the arms race. Is there an independent public assessment of the intelligence data? No. Why not? Because the data are secret. So we have here a machine that goes by itself, a kind of de facto conspiracy to prevent tensions from falling below a minimum level of bureaucratic acceptability.

It is evident that many national institutions and dogmas, however effective they may once have been, are now in need of change. No nation is yet well-fitted to the world of the twenty-first century. The challenge then is not in selective glorification of the past, or in defending the national icons, but in devising a path that will carry us through a time of great mutual peril. To accomplish this, we need all the help we can get.

A central lesson of science is that to understand complex issues (or even simple ones), we must try to free our minds of dogma and to guarantee the freedom to publish, to contradict, and to experiment. Arguments from authority are unacceptable. We are all fallible, even leaders. But however clear it is that criticism is necessary for progress, governments tend to resist. The ultimate example is Hitler's Germany. Here is an excerpt from a speech by the Nazi Party leader Rudolf Hess on June 30, 1934: "One man remains beyond all criticism, and that is the Führer. This is because everyone senses and knows: He is always right, and he will always be right. The National Socialism of all of us is anchored in uncritical loyalty, in a surrender to the Führer."

The convenience of such a doctrine for national leaders is further clarified by Hitler's remark: "What good fortune for those in power that people do not think!" Widespread intellectual and moral docility may be convenient for leaders in the short term, but it is suicidal for nations in the long term. One of the criteria for national leadership should therefore be a talent for understanding, encouraging, and making constructive use of vigorous criticism.

So when those who once were silenced and humiliated by state terror now are able to speak out — fledgling civil libertarians flexing their wings — of course they find it exhilarating, and so does any lover of freedom who witnesses it. *Glasnost* and *perestroika* exhibit to the rest of the world the human scope of Soviet society that past policies have masked. They provide error-correcting mechanisms at all levels of Soviet society. They are essential for economic well-being. They permit real improvements in international cooperation and a major reversal of the nuclear arms race. *Glasnost* and *perestroika* are thus good for the Soviet Union and good for the United States.

There is, of course, opposition to *glasnost* and *perestroika* in the Soviet Union: by those who must now demonstrate their abilities competitively rather than sleepwalking through lifetime tenure; by those unaccustomed to the responsibilities of democracy; by those in no mood, after decades of following the norms, to be taken to task for past behavior. And in the United States too, there are those who oppose *glasnost* and *perestroika:* Some argue it is a trick to lull the West, while the Soviet Union gathers its strength to emerge as a still more formidable rival. Some prefer the old kind of Soviet Union — debilitated by its lack of democracy, easily demonized, readily caricatured. (Americans, complacent about their own democratic forms for too long, have something to learn from *glasnost* and *perestroika* as well. This by itself makes some Americans uneasy.) With such powerful forces arrayed for and against reform, no one can know the outcome.

In both countries, what passes for public debate is still, on closer examination, mainly repetition of national slogans, appeal to popular prejudice, innuendo, self-justification, misdirection, incantation of homilies when evidence is asked for, and a thorough contempt for the intelligence of the citizenry. What we need is an admission of how little we actually know about how to pass safely through the next few decades, the courage to examine a wide range of alternative programs and, most of all, a dedication not to dogma but to solutions. Finding any solution will be hard enough. Finding ones that perfectly correspond to eighteenth- or nineteenth-century political doctrines will be much more difficult.

Our two nations must help one another figure out what changes must be made; the changes must help both sides; and our perspective must embrace a future beyond the next Presidential term of office or the next Five Year Plan. We need to reduce military budgets; raise living standards; engender respect for learning; support science, scholarship, invention, and industry; promote free inquiry; reduce domestic coercion; involve the workers more in managerial decisions; and promote a genuine respect and understanding derived from an acknowledgment of our common humanity and our common jeopardy.

Although we must cooperate to an unprecedented degree, I am not arguing against healthy competition. But let us compete in finding ways to reverse the nuclear arms race and to make massive reductions in conventional forces; in eliminating government corruption; in making most of the world agriculturally self-sufficient. Let us vie in art and science, in music and literature, in technological innovation. Let us have an honesty race. Let us compete in relieving suffering and ignorance and disease; in respecting national independence worldwide; in formulating and implementing an ethic for responsible stewardship of the planet.

Let us learn from one another. Capitalism and socialism have been mutually borrowing methods and doctrine in largely unacknowledged plagiarisms for a century. Neither the U.S. nor the Soviet Union has a monopoly on truth and virtue. I would like to see us compete in cooperativeness. In the 1970s, apart from treaties constraining the nuclear arms race, we had some notable successes in working together — the elimination of smallpox worldwide, efforts to prevent South African nuclear weapons development, the *Apollo-Soyuz* joint manned spaceflight. We can now do much better. Let us begin with a few joint projects of great scope and vision — in relief of starvation, especially in nations such as Ethiopia, which are victimized by superpower rivalry; in identifying and defusing long-term environmental catastrophes that are products of our technology; in fusion physics to provide a safe energy source for the future; in joint exploration of Mars, culminating in the first landing of human beings — Soviets and Americans — on another planet.

Perhaps we will destroy ourselves. Perhaps the common enemy within us will be too strong for us to recognize and overcome. Perhaps the world will be reduced to medieval conditions or far worse.

But I have hope. Lately there are signs of change — tentative but in the right direction and, by previous standards of national behavior, swift. Is it possible that we — we Americans, we Soviets, we humans — are at last coming to our senses and beginning to work together on behalf of the species and the planet?

Nothing is promised. History has placed this burden on our shoulders. It is up to us to build a future worthy of our children and grandchildren.

THE CENSORSHIP

Here in chronological order, keyed to the sequence of paragraphs, are some of the more egregious or interesting changes inflicted on the article as it appeared in *Ogon-*

yok. The censored material is shown here in boldface, ordinary type indicates excerpts from the original article, and bracketed italic type, comments by me.

¶ 3. **. . . that lie at the base of a poorly understood food chain — at the top of which precariously teeter we.** *[Without this phrase, the danger of ozone depletion seems much less.]*

¶ 4. . . . enough nuclear weapons each year to destroy **every sizable city on the planet.** *[The last six words are replaced by* ***any city.*** *But this defocus from the number of bombs produced each year to the power of a single bomb minimizes the nuclear threat.]*

¶ 4. **. . . in an already burdened national leader.** *[Does it diminish confidence in the government to think that the leader may be burdened?]*

¶ 4. **. . . intimidation and** war.

¶ 7. **. . . wounded pride and** professed moral rectitude.

¶ 7. . . . hatred and fear **intentionally fanned by the respective agencies of national propaganda . . .**

¶ 8. **In 1899, two years before becoming President, Theodore** Roosevelt . . . *[This seems especially nasty, because the material removed makes it likely that 99 per-*

cent of Soviet readers will think it's Franklin and not Theodore Roosevelt being quoted.]

¶ 8. **This is not just a matter of mean-spirited Soviet propaganda.**

¶ 9. **. . . July 2 . . .**

¶ 9. **. . . the secret protocol to his nonaggression pact with Hitler . . .**

¶ 9. **. . . and how many millions more were killed in consequence.**

¶11. **. . . the deficiencies of the argument would be better understood, however, had the Soviet Union not been in the habit of gobbling up other countries.**

¶18. So when those who once were silenced and humiliated **by state terror** now are able to speak out — **fledgling civil libertarians flexing their wings** — of course they find it exhilarating, **and so does any lover of freedom who witnesses it.**

¶19. **. . . readily caricatured . . .**

¶20. **In both countries, what passes for public debate is still, on closer examination, mainly repetition of national slogans, appeal to popular prejudice, innuendo, self-justification, misdirection, incantation of homilies when evidence is asked for, and a thorough contempt for the intelligence of the citizenry.**

¶20. Finding any solution will be hard enough. **Finding ones that perfectly correspond to 18th- or 19th-century political doctrines will be much more difficult.** *[Marxism is, of course, a 19th-century political and economic doctrine.]*

¶23. . . . **in largely unacknowledged plagiarisms** for a century. Neither the U.S. nor the Soviet Union has a monopoly on truth and virtue.

¶26. **Nothing is promised.** *[It is one of the self-congratulatory but unscientific tenets of orthodox Marxism that the ultimate triumph of Communism is foreordained by unseen historical forces.]*

The biggest Soviet concern was the quotation from Lenin (and by implication from Tukhachevsky) in Paragraph 9. After repeated requests, which I refused, for me to remove the material, the *Ogonyok* article made a point of including the following footnote: "The editorial staff of *Ogonyok* consulted the relevant archives. However, neither this quotation nor any other similar statement of V. I. Lenin turned up. We regret that the millions of readers of the magazine *Parade* will be misled by this quotation, on the basis of which Carl Sagan has built his conclusions." This provided, it seemed to me, a somewhat sour note.

But time passed, new archives were opened, revised histories became available and acceptable, Lenin was demythologized, and the situation resolved itself. In Arbatov's own memoirs appears the following gracious note:

> Here I have an apology to make. In my comments in *Ogonyok* in 1988, discussing an article by the astronomer Carl Sagan, I brushed aside his conclusion that Tukhachevsky's Polish campaign had been an attempt at exporting revolution. This was due to the usual defensiveness, which became a conditioned reflex, and the fact that we got into the habit over many years (eventually it became second nature) of sweeping "inconvenient" facts under the rug. I, for example, have only recently studied these pages of our history with any degree of care.

Chapter 15

ABORTION: IS IT POSSIBLE TO BE BOTH "PRO-LIFE" AND "PRO-CHOICE"?*

Mankind likes to think in terms of extreme opposites. It is given to formulating its beliefs in terms of Either-Ors, between which it recognizes no intermediate possibilities. When forced to recognize that the extremes cannot be acted upon, it is still inclined to hold that they are all right in theory, but that when it comes to practical matters circumstances compel us to compromise.

JOHN DEWEY,
Experience and Education, I (1938)

*Cowritten with Ann Druyan and published first in *Parade* magazine as "The Question of Abortion: A Search for Answers," April 22, 1990.

The issue had been decided years ago. The court had chosen the middle ground. You'd think the fight was over. Instead, there are mass rallies, bombings and intimidation, murders of workers at abortion clinics, arrests, intense lobbying, legislative drama, Congressional hearings, Supreme Court decisions, major political parties almost defining themselves on the issue, and clerics threatening politicians with perdition. Partisans fling accusations of hypocrisy and murder. The intent of the Constitution and the will of God are equally invoked. Doubtful arguments are trotted out as certitudes. The contending factions call on science to bolster their positions. Families are divided, husbands and wives agree not to discuss it, old friends are no longer speaking. Politicians check the latest polls to discover the dictates of their consciences. Amid all the shouting, it is hard for the adversaries to hear one another. Opinions are polarized. Minds are closed.

Is it wrong to abort a pregnancy? Always? Sometimes? Never? How do we decide? We wrote this article to understand better what the contending views are and to see if we ourselves could find a position that would satisfy us both. Is there no middle ground? We had to weigh the arguments of both sides for consistency and to pose test cases, some of which are purely hypothetical. If in some of these tests we seem to go too far, we ask the reader to be patient with us — we're trying to stress the various positions to the breaking point to see their weaknesses and where they fail.

In contemplative moments, nearly everyone recognizes that the issue is not wholly one-sided. Many partisans of differing views, we find, feel some disquiet, some unease when confronting what's behind the opposing arguments. (This is partly why such confrontations are avoided.) And the issue surely touches on deep questions: What are our responsibilities to one another? Should we permit the state to intrude into the most intimate and personal aspects of our lives? Where are the boundaries of freedom? What does it mean to be human?

Of the many actual points of view, it is widely held — especially in the media, which rarely have the time or the inclination to make fine distinctions — that there are only two: "pro-choice" and "pro-life." This is what the two principal warring camps like to call themselves, and that's what we'll call them here. In the simplest characterization, a pro-choicer would hold that the decision to abort a pregnancy is to be made only by the woman; the state has no right to interfere. And a pro-lifer would hold that, from the moment of conception, the embryo or fetus is alive; that this life imposes on us a moral obligation to preserve it; and that abortion is tantamount to murder. Both names — pro-choice and pro-life — were picked with an eye toward influencing those whose minds are not yet made up: Few people wish to be counted either as being against freedom of choice or as opposed to life. Indeed, freedom and life are two of our most cherished values, and here they seem to be in fundamental conflict.

Let's consider these two absolutist positions in turn. A newborn baby is surely the same being it was just before birth. There is good evidence that a late-term fetus responds to sound — including music, but especially its mother's voice. It can suck its thumb or do a somersault. Occasionally, it generates adult brain-wave patterns. Some people claim to remember being born, or even the uterine environment. Perhaps there is thought in the womb. It's hard to maintain that a transformation to full personhood happens abruptly at the moment of birth. Why, then, should it be murder to kill an infant the day after it was born but not the day before?

As a practical matter, this isn't very important: Less than 1 percent of all tabulated abortions in the United States are listed in the last three months of pregnancy (and, on closer investigation, most such reports turn out to be due to miscarriage or miscalculation). But third-trimester abortions provide a test of the limits of the pro-choice point of view. Does a woman's "innate right to control her own body" encompass the right to kill a near-term fetus who is, for all intents and purposes, identical to a newborn child?

We believe that many supporters of reproductive freedom are troubled at least occasionally by this question. But they are reluctant to raise it because it is the beginning of a slippery slope. If it is impermissible to abort a pregnancy in the ninth month, what about the eighth, seventh, sixth . . . ? Once we acknowledge that the state can interfere at *any* time in the pregnancy, doesn't it follow that the state can interfere at all times?

This conjures up the specter of predominantly male, predominantly affluent legislators telling poor women they must bear and raise alone children they cannot afford to bring up; forcing teenagers to bear children they are not emotionally prepared to deal with; saying to women who wish for a career that they must give up their dreams, stay home, and bring up babies; and, worst of all, condemning victims of rape and incest to carry and nurture the offspring of their assailants.* Legislative prohibitions on abortion arouse the suspicion that their real intent is to control the independence and sexuality of women. Why should legislators have any right at all to tell women what to do with their bodies? To be deprived of reproductive freedom is demeaning. Women are fed up with being pushed around.

And yet, by consensus, all of us think it proper that there be prohibitions against, and penalties exacted for, murder. It would be a flimsy defense if the murderer pleads that this is just between him and his victim and none of the government's business. If killing a fetus is truly killing a human being,

*Two of the most energetic pro-lifers of all time were Hitler and Stalin — who immediately upon taking power criminalized previously legal abortions. Mussolini, Ceauşescu, and countless other nationalist dictators and tyrants have done likewise. Of course, this is not by itself a pro-choice argument, but it does alert us to the possibility that being against abortion may not always be part of a deep commitment to human life.

is it not the *duty* of the state to prevent it? Indeed, one of the chief functions of government is to protect the weak from the strong.

If we do not oppose abortion at *some* stage of pregnancy, is there not a danger of dismissing an entire category of human beings as unworthy of our protection and respect? And isn't that dismissal the hallmark of sexism, racism, nationalism, and religious fanaticism? Shouldn't those dedicated to fighting such injustices be scrupulously careful not to embrace another?

There is no right to life in any society on Earth today, nor has there been at any former time (with a few rare exceptions, such as among the Jains of India): We raise farm animals for slaughter; destroy forests; pollute rivers and lakes until no fish can live there; kill deer and elk for sport, leopards for their pelts, and whales for fertilizer; entrap dolphins, gasping and writhing, in great tuna nets; club seal pups to death; and render a species extinct every day. All these beasts and vegetables are as alive as we. What is (allegedly) protected is not life, but *human* life.

And even with that protection, casual murder is an urban commonplace, and we wage "conventional" wars with tolls so terrible that we are, most of us, afraid to consider them very deeply. (Tellingly, state-organized mass murders are often justified by redefining our opponents — by race, nationality, religion, or ideology — as less than human.) That protection, that right to life, eludes the 40,000 children under five who die on our planet each day from preventable starvation, dehydration, disease, and neglect.

Those who assert a "right to life" are for (at most) not just any kind of life, but for — particularly and uniquely — human life. So they too, like pro-choicers, must decide what distinguishes a human being from other animals and when, during gestation, the uniquely human qualities — whatever they are — emerge.

Despite many claims to the contrary, life does not begin at conception: It is an unbroken chain that stretches back nearly to the origin of the Earth, 4.6 billion years ago. Nor does *human* life begin at conception: It is an unbroken chain dating back to the origin of our species, hundreds of thousands of years ago. Every human sperm and egg is, beyond the shadow of a doubt, alive. They are not human beings, of course. However, it could be argued that neither is a fertilized egg.

In some animals, an egg develops into a healthy adult without benefit of a sperm cell. But not, so far as we know, among humans. A sperm and an unfertilized egg jointly comprise the full genetic blueprint for a human being. Under certain circumstances, after fertilization, they can develop into a baby. But most fertilized eggs are spontaneously miscarried. Development into a baby is by no means guaranteed. Neither a sperm and egg separately, nor a fertilized egg, is more than a *potential* baby or a *potential* adult. So if a sperm and egg are as human as the fertilized egg produced by their union, and if it is murder to destroy a fertilized egg — despite the fact that it's only *potentially* a baby — why isn't it murder to destroy a sperm or an egg?

Hundreds of millions of sperm cells (top speed with tails lashing: five inches per hour) are produced in an average human ejaculation. A healthy young man can produce in a week or two enough spermatozoa to double the human population of the Earth. So is masturbation mass murder? How about nocturnal emissions or just plain sex? When the unfertilized egg is expelled each month, has someone died? Should we mourn all those spontaneous miscarriages? Many lower animals can be grown in a laboratory from a single body cell. Human cells can be cloned (perhaps the most famous being the HeLa clone, named after the donor, Helen Lane). In light of such cloning technology, would we be committing mass murder by destroying any potentially clonable cells? By shedding a drop of blood?

All human sperm and eggs are genetic halves of "potential" human beings. Should heroic efforts be made to save and preserve all of them, everywhere, because of this "potential"? Is failure to do so immoral or criminal? Of course, there's a difference between taking a life and failing to save it. And there's a big difference between the probability of survival of a sperm cell and that of a fertilized egg. But the absurdity of a corps of high-minded semen-preservers moves us to wonder whether a fertilized egg's mere "potential" to become a baby really does make destroying it murder.

Opponents of abortion worry that, once abortion is permissible immediately after conception, no argument will restrict it at any later time in the pregnancy. Then, they fear, one day it will be permissible to murder a fetus that is unambiguously a human being. Both pro-choicers and pro-lifers (at least some of them) are pushed toward absolutist positions by parallel fears of the slippery slope.

Another slippery slope is reached by those pro-lifers who are willing to make an exception in the agonizing case of a pregnancy resulting from rape or incest. But why should the right to live depend on the *circumstances* of conception? If the same child were to result, can the state ordain life for the offspring of a lawful union but death for one conceived by force or coercion? How can this be just? And if exceptions are extended to such a fetus, why should they be withheld from any other fetus? This is part of the reason some pro-lifers adopt what many others consider the outrageous posture of opposing abortions under any and all circumstances — only excepting, perhaps, when the life of the mother is in danger.*

By far the most common reason for abortion worldwide is birth control. So shouldn't opponents of abortion be handing

*Martin Luther, the founder of Protestantism, opposed even this exception: "If they become tired or even die through bearing children, that does not matter. Let them die through fruitfulness — that is why they are there" (Luther, *Vom Ebelichen Leben* [1522]).

out contraceptives and teaching school children how to use them? That would be an effective way to reduce the number of abortions. Instead, the United States is far behind other nations in the development of safe and effective methods of birth control — and, in many cases, opposition to such research (and to sex education) has come from the same people who oppose abortions.*

The attempt to find an ethically sound and unambiguous judgment on when, if ever, abortion is permissible has deep historical roots. Often, especially in Christian tradition, such attempts were connected with the question of when the soul enters the body — a matter not readily amenable to scientific investigation and an issue of controversy even among learned theologians. Ensoulment has been asserted to occur in

*Likewise, shouldn't pro-lifers count birthdays from the moment of conception, and not from the moment of birth? Shouldn't they closely interrogate their parents on their sexual history? There would of course be some irreducible uncertainty: It may take hours or days after the sex act for conception to happen (a particular difficulty for pro-lifers who also wish to dally with Sun-sign astrology).

the sperm before conception, at conception, at the time of "quickening" (when the mother is first able to feel the fetus stirring within her), and at birth. Or even later.

Different religions have different teachings. Among hunter-gatherers, there are usually no prohibitions against abortion, and it was common in ancient Greece and Rome. In contrast, the more severe Assyrians impaled women on stakes for attempting abortion. The Jewish Talmud teaches that the fetus is not a person and has no rights. The Old and New Testaments — rich in astonishingly detailed prohibitions on dress, diet, and permissible words — contain not a word specifically prohibiting abortion. The only passage that's remotely relevant (Exodus 21:22) decrees that if there's a fight and a woman bystander should accidentally be injured and made to miscarry, the assailant must pay a fine.

Neither St. Augustine nor St. Thomas Aquinas considered early-term abortion to be homicide (the latter on the grounds that the embryo doesn't *look* human). This view was embraced by the Church in the Council of Vienne in 1312, and has never been repudiated. The Catholic Church's first and long-standing collection of canon law (according to the leading historian of the

Church's teaching on abortion, John Connery, S.J.) held that abortion was homicide only after the fetus was already "formed" — roughly, the end of the first trimester.

But when sperm cells were examined in the seventeenth century by the first microscopes, they were thought to show a fully formed human being. An old idea of the homunculus was resuscitated — in which within each sperm cell was a fully formed tiny human, within whose testes were innumerable other homunculi, etc., *ad infinitum.* In part through this misinterpretation of scientific data, in 1869 abortion at any time for any reason became grounds for excommunication. It is surprising to most Catholics and others to discover that the date was not much earlier.

From colonial times to the nineteenth century, the choice in the United States was the woman's until "quickening." An abortion in the first or even second trimester was at worst a misdemeanor. Convictions were rarely sought and almost impossible to obtain, because they depended entirely on the woman's own testimony of whether she had felt quickening, and because of the jury's distaste for prosecuting a woman for exercising her right to choose. In 1800 there was not, so far as is known, a single statute in the United States concerning abortion. Advertisements for drugs to induce abortion could be found in virtually every newspaper and even in many church publications — although the language used was suitably euphemistic, if widely understood.

But by 1900, abortion had been banned at *any* time in pregnancy by every state in the Union, except when necessary to save the woman's life. What happened to bring about so striking a reversal? Religion had little to do with it. Drastic economic and social conversions were turning this country from an agrarian to an urban-industrial society. America was in the process of changing from having one of the highest birthrates in the world to one of the lowest. Abortion certainly played a role and stimulated forces to suppress it.

One of the most significant of these forces was the medical profession. Up to the mid-nineteenth century, medicine was an uncertified, unsupervised business. Anyone could hang up a shingle and call himself (or herself) a doctor. With the rise of a new, university-educated medical elite, anxious to enhance the status and influence of physicians, the American Medical Association was formed. In its first decade, the AMA began lobbying against abortions performed by anyone except licensed physicians. New knowledge of embryology, the physicians said, had shown the fetus to be human even before quickening.

Their assault on abortion was motivated not by concern for the health of the woman but, they claimed, for the welfare of the fetus. You had to be a physician to know when abortion was morally justified, because the question depended on scientific and medical facts understood only by physicians. At the same time, women were effectively excluded from the medical schools, where such arcane knowledge could be acquired. So, as things worked out, women had almost nothing to say about terminating their own pregnancies. It was also up to the physician to decide if the pregnancy posed a threat to the woman, and it was entirely at his discretion to determine what was and was not a threat. For the rich woman, the threat might be a threat to her emotional tranquillity or even to her lifestyle. The poor woman was often forced to resort to the back alley or the coat hanger.

This was the law until the 1960s, when a coalition of individuals and organizations, the AMA now among them, sought to overturn it and to reinstate the more traditional values that were to be embodied in *Roe* v. *Wade.*

* * *

If you deliberately kill a human being, it's called murder. If you deliberately kill a chimpanzee — biologically, our closest relative, sharing 99.6 percent of our active genes — whatever else it is, it's not murder. To date, murder uniquely applies to killing human beings. Therefore, the question of when personhood (or, if we like, ensoulment) arises is key to the abortion debate. When does the fetus become human? When do distinct and characteristic human qualities emerge?

We recognize that specifying a precise moment will overlook individual differences. Therefore, if we must draw a line, it ought to be drawn conservatively — that is, on the early side. There are people who object to having to set some numerical limit, and we share their disquiet; but if there is to be a law on this matter, and it is to effect some useful compromise between the two absolutist positions, it must specify, at least roughly, a time of transition to personhood.

Every one of us began from a dot. A fertilized egg is roughly the size of the period at the end of this sentence. The momentous meeting of sperm and egg generally occurs in one of the two fallopian tubes. One cell becomes two, two become four, and so on

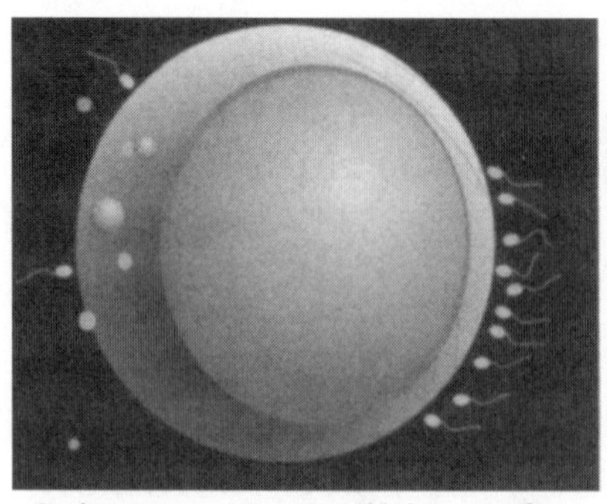

A human egg cell just after fertilization, partially surrounded by the runner-up sperm cells. The roughly 300 million also-rans have not yet arrived.

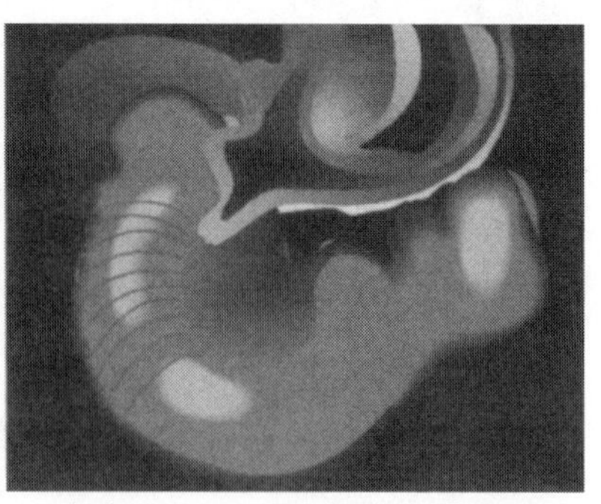

A human embryo three weeks after conception, about the size of a pencil point, with head at the right. The egmentation extending to the tail resembles that of a worm.

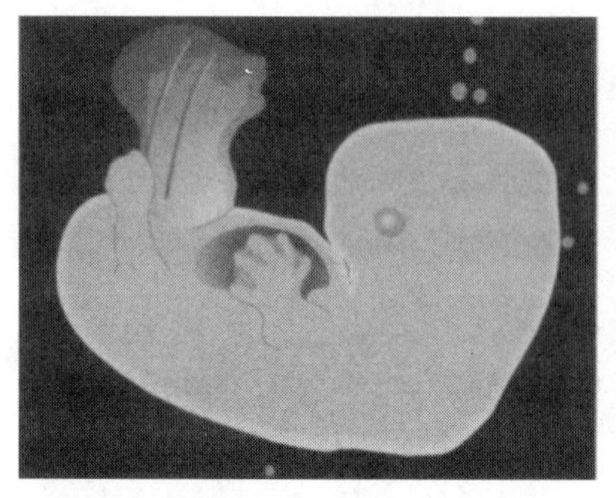

A human embryo at the end of the fifth week after conception. The tail is curled under the leg buds. The face, seen here in profile, has a distinctly reptilian aspect.

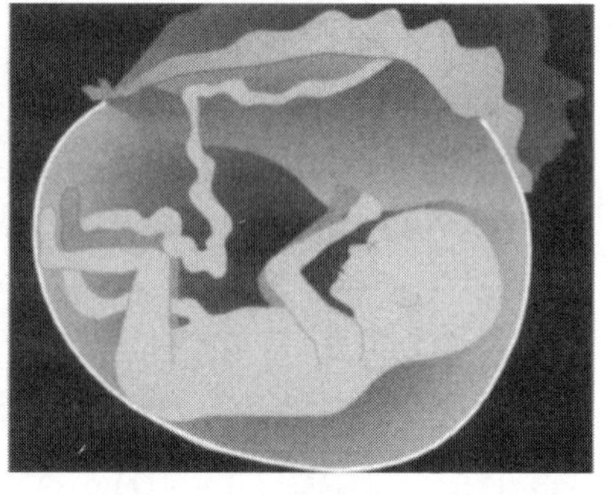

A 16-week-old fetus seems quite human externally. But it cannot yet move on its own enough to be detectable as "quickening," nor can it survive outside the womb.

Drawings based on photographs by Lennart Nilsson/Bonnier Alba AB

— an exponentiation of base-2 arithmetic. By the tenth day the fertilized egg has become a kind of hollow sphere wandering off to another realm: the womb. It destroys tissue in its path. It sucks blood from capillaries. It bathes itself in maternal blood, from which it extracts oxygen and nutrients. It establishes itself as a kind of parasite on the walls of the uterus.

- By the third week, around the time of the first missed menstrual period, the forming embryo is about 2 millimeters long and is developing various body parts. Only at this stage does it begin to be dependent on a rudimentary placenta. It looks a little like a segmented worm.*

*A number of right-wing and Christian fundamentalist publications have criticized this argument — on the grounds that it is based on an obsolete doctrine, called recapitulation, of a nineteenth-century German biologist. Ernst Haeckel proposed that the steps in the individual embryonic development of an animal retrace (or "recapitulate") the stages of evolutionary development of its ancestors. Recapitulation has been exhaustively and skeptically treated by the evolutionary biologist Stephen Jay Gould (in his book *Ontogeny and Phylogeny* [Cambridge, Mass.: Harvard University

- By the end of the fourth week, it's about 5 millimeters (about 1/5 inch) long. It's recognizable now as a vertebrate, its tube-shaped heart is beginning to beat, something like the gill arches of a fish or an amphibian become conspicuous, and there is a pronounced tail. It looks rather like a newt or a tadpole. This is the end of the first month after conception.
- By the fifth week, the gross divisions of the brain can be distinguished. What will later develop into eyes are apparent, and little buds appear — on their way to becoming arms and legs.
- By the sixth week, the embryo is 13 millimeters (about 1/2 inch) long. The eyes are still on the side of the head, as in most animals, and the reptilian face has connected slits where the mouth and nose eventually will be.

Press, 1977]). But our article offered not a word about recapitulation, as the reader of this chapter may judge. The comparisons of the human fetus with other (adult) animals is based on the appearance of the fetus (see illustrations). Its nonhuman form, and nothing about its evolutionary history, is the key to the argument of these pages.

- By the end of the seventh week, the tail is almost gone, and sexual characteristics can be discerned (although both sexes look female). The face is mammalian but somewhat piglike.
- By the end of the eighth week, the face resembles that of a primate but is still not quite human. Most of the human body parts are present in their essentials. Some lower brain anatomy is well-developed. The fetus shows some reflex response to delicate stimulation.
- By the tenth week, the face has an unmistakably human cast. It is beginning to be possible to distinguish males from females. Nails and major bone structures are not apparent until the third month.
- By the fourth month, you can tell the face of one fetus from that of another. Quickening is most commonly felt in the fifth month. The bronchioles of the lungs do not begin developing until approximately the sixth month, the alveoli still later.

So, if only a person can be murdered, when does the fetus attain personhood? When its face becomes distinctly human, near the end of the first trimester? When the fetus becomes responsive to stimuli — again, at the end of the first trimester? When it becomes active enough to be felt as quickening, typically in the middle of the second trimester? When the lungs have reached a stage of development sufficient that the fetus might, just conceivably, be able to breathe on its own in the outside air?

The trouble with these particular developmental milestones is not just that they're arbitrary. More troubling is the fact that none of them involves *uniquely human* char-

THE FIRST EIGHT WEEKS

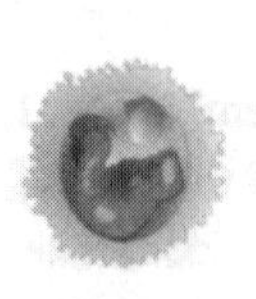

Stages in the development of the embryo and fetus during the first eight weeks after conception. Shown at far left is the newly fertilized egg, containing 46 chromosomes — the full genetic blueprint, half contributed by the sperm, half by the egg. Each successive illustration is one week further along in the pregnancy, except for the last, which corresponds to the eighth week. After stages

acteristics — apart from the superficial matter of facial appearance. All animals respond to stimuli and move of their own volition. Large numbers are able to breathe. But that doesn't stop us from slaughtering them by the billions. Reflexes and motion and respiration are not what make us human.

Other animals have advantages over us — in speed, strength, endurance, climbing or burrowing skills, camouflage, sight or smell or hearing, mastery of the air or water. Our one great advantage, the secret of our success, is thought — characteristically human thought. We are able to think

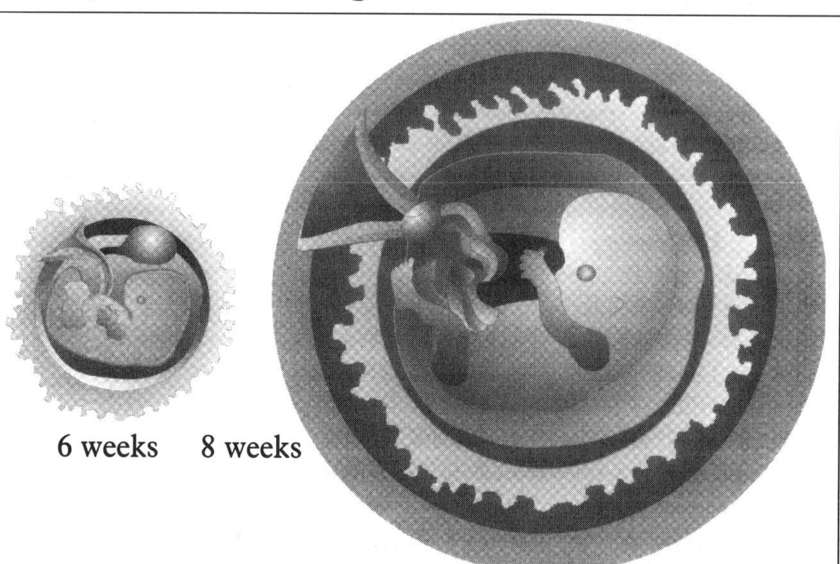
6 weeks 8 weeks

resembling a worm, an amphibian, a reptile, and a lower mammal, in the eighth week recognizable primate (monkey, ape, human) features appear. It will be many more months before lungs develop and distinctly human brain activity commences.

things through, imagine events yet to occur, figure things out. That's how we invented agriculture and civilization. Thought is our blessing and our curse, and it makes us who we are.

Thinking occurs, of course, in the brain — principally in the top layers of the convoluted "gray matter" called the cerebral cortex. The roughly 100 billion neurons in the brain constitute the material basis of thought. The neurons are connected to each other, and their linkups play a major role in what we experience as thinking. But large-scale linking up of neurons doesn't begin until the 24th to 27th week of pregnancy — the sixth month.

By placing harmless electrodes on a subject's head, scientists can measure the electrical activity produced by the network of neurons inside the skull. Different kinds of mental activity show different kinds of brain waves. But brain waves with regular patterns typical of adult human brains do not appear in the fetus until about the 30th week of pregnancy — near the beginning of the third trimester. Fetuses younger than this — however alive and active they may be — lack the necessary brain architecture. They cannot yet think.

Acquiescing in the killing of any living

creature, especially one that might later become a baby, is troublesome and painful. But we've rejected the extremes of "always" and "never," and this puts us — like it or not — on the slippery slope. If we are forced to choose a developmental criterion, then this is where we draw the line: when the beginning of characteristically human thinking becomes barely possible.

It is, in fact, a very conservative definition: Regular brain waves are rarely found in fetuses. More research would help. (Well-defined brain waves in fetal baboons and fetal sheep also begin only late in gestation.) If we wanted to make the criterion still more stringent, to allow for occasional precocious fetal brain development, we might draw the line at six months. This, it so happens, is where the Supreme Court drew it in 1973 — although for completely different reasons.

Its decision in the case of *Roe* v. *Wade* changed American law on abortion. It permits abortion at the request of the woman without restriction in the first trimester and, with some restrictions intended to protect her health, in the second trimester. It allows states to forbid abortion in the third trimester, except when there's a serious threat to the life or health of the woman. In the 1989

Webster decision, the Supreme Court declined explicitly to overturn *Roe* v. *Wade* but in effect invited the 50 state legislatures to decide for themselves.

What was the reasoning in *Roe* v. *Wade*? There was no legal weight given to what happens to the children once they are born, or to the family. Instead, a woman's right to reproductive freedom is protected, the court ruled, by constitutional guarantees of privacy. But that right is not unqualified. The woman's guarantee of privacy and the fetus's right to life must be weighed — and when the court did the weighing, priority was given to privacy in the first trimester and to life in the third. The transition was decided not from any of the considerations we have been dealing with so far in this chapter — not when "ensoulment" occurs, not when the fetus takes on sufficient human characteristics to be protected by laws against murder. Instead, the criterion adopted was whether the fetus could live outside the mother. This is called "viability" and depends in part on the ability to breathe. The lungs are simply not developed, and the fetus cannot breathe — no matter how advanced an artificial lung it might be placed in — until about the 24th week, near the start of the sixth month. This is why *Roe* v. *Wade* permits the states to prohibit abortions in the last trimester. It's a very pragmatic criterion.

If the fetus at a certain stage of gestation would be viable outside the womb, the argument goes, then the right of the fetus to life overrides the right of the woman to privacy. But just what does "viable" mean? Even a full-term newborn is not viable without a great deal of care and love. There was a time before incubators, only a few decades ago, when babies in their seventh month were unlikely to be viable. Would aborting in the seventh month have been permissible then? After the invention of incubators, did aborting pregnancies in the seventh month suddenly become immoral? What happens if, in the future, a new technology develops so that an artificial womb can sustain a fetus even before the sixth month by delivering oxygen and nutrients through the blood — as the mother does through the placenta and into the fetal blood system? We grant that this technology is unlikely to be developed soon or become available to many. But *if* it were available, does it then become immoral to abort earlier than the sixth month, when previously it was moral? A morality that depends on, and changes with, technology is a fragile morality; for some, it is also an unacceptable morality.

And why, exactly, should breathing (or kidney function, or the ability to resist disease) justify legal protection? If a fetus can be shown to think and feel but not be able to breathe, would it be all right to kill it? Do we value breathing more than thinking and feeling? Viability arguments cannot, it seems to us, coherently determine when abortions are permissible. Some other criterion is needed. Again, we offer for consideration the earliest onset of human thinking as that criterion.

Since, on average, fetal thinking occurs even later than fetal lung development, we find *Roe* v. *Wade* to be a good and prudent decision addressing a complex and difficult issue. With prohibitions on abortion in the last trimester — except in cases of grave medical necessity — it strikes a fair balance between the conflicting claims of freedom and life.

When this article appeared in *Parade* it was accompanied by a box giving a 900 telephone number for the readers to express their views on the abortion issue. An astonishing 380,000 people called in. They were able to express the following four options: "Abortion after the instant of conception is murder," "A woman has the right to choose abortion any time during pregnancy," "Abortion should be permitted within the first three months of pregnancy," and "Abortion should be permitted within the first six months of pregnancy." *Parade* is published on Sunday, and by Monday, opinions were well divided among these four options. Then Mr. Pat Robertson, a Christian fundamentalist evangelist and 1992 Republican Presidential candidate, appeared Monday on his regularly scheduled daily television program, urged his followers to pull *Parade* "out of the garbage" and send back the clear message that killing a human zygote is murder. They did. The generally pro-choice attitude of most Americans — as repeatedly shown in demographically controlled opinion polls, and as had been reflected by the early 900 number results — was overwhelmed by political organization.

Chapter 16

THE RULES OF THE GAME

Everything morally right derives from one of four sources: it concerns either full perception or intelligent development of what is true; or the preservation of organized society, where every man is rendered his due and all obligations are faithfully discharged; or the greatness and strength of a noble, invincible spirit; or order and moderation in everything said and done, whereby is temperance and self-control.

CICERO,
De Officiis, I, 5 (45–44 B.C.)

I remember the end of a long ago perfect day in 1939 — a day that powerfully influenced my thinking, a day when my parents introduced me to the wonders of the New York World's Fair. It was late, well past my bedtime. Safely perched on my father's shoulders, holding onto his ears, my mother reassuringly at my side, I turned to see the great Trylon and Perisphere, the architectural icons of the Fair, illuminated in shimmering blue pastels. We were abandoning the future, the "World of Tomorrow," for the BMT subway train. As we paused to rearrange our possessions, my father got to talking with a small, tired man carrying a tray around his neck. He was selling pencils. My father reached into the crumpled brown paper bag that held the remains of our lunches, withdrew an apple, and handed it to the pencil man. I let out a loud wail. I disliked apples then, and had refused this one both at lunch and at dinner. But I had, nevertheless, a proprietary interest in it. It was my *apple, and my father had just given it away to a funny-looking stranger — who, to compound my anguish, was now glaring unsympathetically in my direction.*

Although my father was a person of nearly limitless patience and tenderness, I could see he was disappointed in me. He swept me up and hugged me tight to him.

"He's a poor stiff, out of work," he said to me, too quietly for the man to hear. "He hasn't eaten all day. We have enough. We can give him an apple."

I reconsidered, stifled my sobs, took another wistful glance at the World of Tomorrow, and gratefully fell asleep in his arms.

Moral codes that seek to regulate human behavior have been with us not only since the dawn of civilization but also among our precivilized, and highly social, hunter-gatherer ancestors. And even earlier. Different societies have different codes. Many cultures say one thing and do another. In a few fortunate societies, an inspired lawgiver lays down a set of rules to live by (and more often than not claims to have been instructed by a god — without which few would follow the prescriptions). For example, the codes of Ashoka (India), Hammurabi (Babylon), Lycurgus (Sparta), and Solon (Athens), which once held sway over mighty civilizations, are today largely defunct. Perhaps they misjudged human nature and asked too much of us. Perhaps experience from one epoch or culture is not wholly applicable to another.

Surprisingly, there are today efforts — tentative but emerging — to approach the matter scientifically; i.e., experimentally.

In our everyday lives as in the momentous relations of nations, we must decide: What does it mean to do the right thing? Should we help a needy stranger? How do we deal with an enemy? Should we ever take advantage of someone who treats us kindly? If hurt by a friend, or helped by an enemy, should we reciprocate in kind; or does the totality of past behavior outweigh any recent departures from the norm?

Examples: Your sister-in-law ignores your snub and invites you over for Christmas dinner; should you accept? Shattering a four-year-long worldwide voluntary moratorium, China resumes nuclear weapons testing; should we? How much should we give to charity? Serbian soldiers systematically rape Bosnian women; should Bosnian soldiers systematically rape Serbian women? After centuries of oppression, the Nationalist Party leader F. W. de Klerk makes overtures to the African National Congress; should Nelson Mandela and the ANC have reciprocated? A coworker makes you look bad in front of the boss; should you try to get even? Should we cheat on our income tax returns? If we can get away with it? If an oil company supports a symphony orchestra or sponsors a refined TV drama, ought we to ignore its pollution of the environment? Should we be kind to aged relatives, even if they drive us nuts? Should you cheat at cards? Or on a larger scale? Should we kill killers?

In making such decisions, we're concerned not only with doing right but also with what works — what makes us and the rest of society happier and more secure. There's a tension between what we call ethical and what we call pragmatic. If, even in the long run, ethical behavior were self-defeating, eventually we would not call it ethical, but foolish. (We might even claim to respect it in principle, but ignore it in practice.) Bearing in mind the variety and complexity of human behavior, are there any simple rules — whether we call them ethical or pragmatic — that actually work?

How do we decide what to do? Our responses are partly determined by our perceived self-interest. We reciprocate in kind or act contrary because we hope it will accomplish what we want. Nations assemble or blow up nuclear weapons so other countries won't trifle with them. We return good for evil because we know that we can thereby sometimes touch people's sense of justice, or shame them into being nice. But sometimes we're not motivated selfishly. Some people seem just naturally kind. We may accept aggravation from aged parents or from children, because we love them and want them to be happy, even if it's at some cost to us. Sometimes we're tough with our children and cause them a little unhappiness, because we want to mold their characters and believe that the long-term results will bring them more happiness than the short-term pain.

Cases are different. Peoples and nations are different. Knowing how to negotiate this labyrinth is part of wisdom. But bearing in mind the variety and complexity of human behavior, are there some simple rules, whether we call them ethical or pragmatic, that actually work? Or maybe we should avoid trying to think it through and just do what feels right. But even then how do we *determine* what "feels right"?

The most admired standard of behavior, in the West at least, is the Golden Rule, attributed to Jesus of Nazareth. Everyone knows its formulation in the first-century Gospel of St. Matthew: **Do unto others as you would have them do unto you.** Almost no one follows it. When the Chinese philosopher Kung-Tzi (known as Confucius in the West) was asked in the fifth century B.C. his opinion of the Golden Rule (by then already well-known), of repaying evil with kindness, he replied, "Then with what will you repay kindness?" Shall the poor woman who envies her neighbor's wealth give what little she has to the rich? Shall the masochist inflict pain on his neighbor? The Golden Rule takes no account of human differences. Are we really capable, after our cheek has been slapped, of turning the other cheek so it too can be slapped? With a heartless adversary, isn't this just a guarantee of more suffering?

The Silver Rule is different: **Do not do unto others what you would not have them do unto you.** It also can be found worldwide, including, a generation before Jesus, in the writings of Rabbi Hillel. The most inspiring twentieth-century exemplars of the Silver Rule were Mohandas Gandhi and Martin Luther King, Jr. They counseled oppressed peoples not to repay violence with violence, but not to be compliant and obedient either. Nonviolent civil disobedience was what they advocated — putting your body on the line, showing, by your willingness to be punished in defying an unjust law, the justice of your cause. They aimed at melting the hearts of their oppressors (and those who had not yet made up their minds).

King paid tribute to Gandhi as the first person in history to convert the Golden or Silver Rules into an effective instrument of social change. And Gandhi made it clear where his approach came from: "I learnt the lesson on nonviolence from my wife, when I tried to bend her to my will. Her determined resistance to my will on the one hand, and her quiet submission to the suffering my stupidity involved on the other, ultimately made me ashamed of myself and cured me of my stupidity in thinking that I was born to rule over her."

Nonviolent civil disobedience has worked notable political change in this century — in prying India loose from British rule and stimulating the end of classic colonialism worldwide, and in providing some civil rights for African-Americans — although the threat of violence by others, however disavowed by Gandhi and King, may have also helped. The African National Congress (ANC) grew up in the Gandhian tradition. But by the 1950s it was clear that nonviolent noncooperation was making no progress whatever with the ruling white Nationalist Party. So in 1961 Nelson Mandela and his colleagues formed the military wing of the ANC, the *Umkhonto we Sizwe*, the Spear of the Nation, on the quite un-Gandhian grounds that the only thing whites understand is force.

Even Gandhi had trouble reconciling the rule of nonviolence with the necessities of defense against those with less lofty rules of conduct: "I have not the qualifications for teaching my philosophy of life. I have barely qualifications for practicing the philosophy I believe. I am but a poor struggling soul yearning to be . . . wholly truthful and wholly nonviolent in thought, word and deed, but ever failing to reach the ideal."

"Repay kindness with kindness," said Confucius, "but evil with justice." This might be called the Brass or Brazen Rule: **Do unto others as they do unto you.** It's the *lex talionis,* "an eye for an eye, and a tooth for a tooth," *plus* "one good turn deserves another." In actual human (and chimpanzee) behavior it's a familiar standard. "If the enemy inclines toward peace, do thou also incline toward peace," President Bill Clinton quoted from the Qur'an at the Israeli-Palestinian peace accords. Without having to appeal to anyone's better nature, we institute a kind of operant conditioning, rewarding them when they're nice to us and punishing them when they're not. We're not pushovers but we're not unforgiving either. It sounds promising. Or is it true that "two wrongs don't make a right"?

Of baser coinage is the Iron Rule: **Do unto others as you like, before they do it unto you.** It is sometimes formulated as "He who has the gold makes the rules," underscoring not just its departure from, but its contempt for the Golden Rule. This is the secret maxim of many, if they can get away with it, and often the unspoken precept of the powerful.

Finally, I should mention two other rules, found throughout the living world. They explain a great deal: One is **Suck up to those above you, and abuse those below.** This is the motto of bullies and the norm in many nonhuman primate societies. It's really the Golden Rule for superiors, the Iron Rule for inferiors. Since there is no known alloy of gold and iron, we'll call it the Tin Rule for its flexibility. The other common rule is **Give precedence in all things to close relatives, and do as you like to others.** This Nepotism Rule is known to evolutionary biologists as "kin selection."

Despite its apparent practicality, there's a fatal flaw in the Brazen Rule: unending vendetta. It hardly matters who starts the violence. Violence begets violence, and each side has reason to hate the other. "There is no way to peace," A. J. Muste said. "Peace *is* the way." But peace is hard and violence is easy. Even if almost everyone is for ending the vendetta, a single act of retribution can stir it up again: A dead relative's sobbing widow and grieving children are before us. Old men and women recall atrocities from their childhoods. The reasonable part of us tries to keep the peace, but the passionate part of us cries out for vengeance. Extremists in the two warring factions can count on one another. They are allied against the rest of us, contemptuous of appeals to understanding and loving-kindness. A few hotheads can force-march a legion of more prudent and rational people to brutality and war.

Many in the West have been so mesmerized by the appalling accords with Adolf Hitler in Munich in 1938 that they are unable to distinguish cooperation and appeasement. Rather than having to judge each gesture and approach on its own merits, we merely decide that the opponent is thoroughly evil, that all his concessions are offered in bad faith, and that force is the only thing he understands. Perhaps for Hitler this was the right judgment. But in general it is not the right judgment, as much as I wish that the invasion of the Rhineland had been forcibly opposed. It consolidates hostility on both sides and makes conflict much more likely. In a world with nuclear weapons, uncompromising hostility carries special and very dire dangers.

Breaking out of a long series of reprisals

is, I claim, very hard. There are ethnic groups who have weakened themselves to the point of extinction because they had no machinery to escape from this cycle, the Kaingáng of the Brazilian highlands, for example. The warring nationalities in the former Yugoslavia, in Rwanda, and elsewhere may provide further examples. The Brazen Rule seems too unforgiving. The Iron Rule promotes the advantage of a ruthless and powerful few against the interests of everybody else. The Golden and Silver Rules seem too complacent. They systematically fail to punish cruelty and exploitation. They hope to coax people from evil to good by showing that kindness is possible. But there are sociopaths who do not much care about the feelings of others, and it is hard to imagine a Hitler or a Stalin being shamed into redemption by good example. Is there a rule between the Golden and the Silver on the one hand and the Brazen, Iron, and Tin on the other which works better than any of them alone?

With so many different rules, how can you tell which to use, which will work? More than one rule may be operating even in the same person or nation. Are we doomed just to guess about this, or to rely on intuition, or just to parrot what we've been taught? Let's try to put aside, just for the moment, whatever rules we've been taught, and those we feel passionately — perhaps from a deeply rooted sense of justice — *must* be right.

Suppose we seek not to confirm or deny what we've been taught, but to find out what really works. Is there a way to *test* competing codes of ethics? Granting that the real world may be much more complicated than any simulation, can we explore the matter scientifically?

We're used to playing games in which somebody wins and somebody loses. Every point made by our opponent puts us that much further behind. "Win-lose" games seem natural, and many people are hard-pressed to think of a game that isn't win-lose. In win-lose games, the losses just balance the wins. That's why they're called "zero-sum" games. There's no ambiguity about your opponent's intentions: Within the rules of the game, he will do anything he can to defeat you.

Many children are aghast the first time they really come face to face with the "lose" side of win-lose games. On the verge of bankruptcy in Monopoly, they plead for special dispensation (forgoing rents, for example), and when this is not forthcoming may, in tears, denounce the game as heartless and unfeeling — which of course it is. (I've seen the board overturned, hotels and "Chance" cards and metal icons spilled onto the floor in spitting anger and humiliation — and not only by children.) Within the rules of Monopoly, there's no way for players to cooperate so that all benefit. That's not how the game is designed. The same is true for boxing, football, hockey, basketball, baseball, lacrosse, tennis, racquetball, chess, all Olympic events, yacht and car racing, pinochle, potsie, and partisan politics. In none of these games is there an opportunity to practice the Golden or Silver Rules, or even the Brazen. There is room only for the Rules of Iron and Tin. If we revere the Golden Rule, why is it so rare in the games we teach our children?

After a million years of intermittently warring tribes we readily enough think in zero-sum mode, and treat every interaction as a contest or conflict. Nuclear war, though (and many conventional wars), economic depression, and assaults on the global environment are all "lose-lose" propositions. Such vital human concerns as love, friendship, parenthood, music, art, and the pursuit of knowledge are "win-win" propositions. Our vision is dangerously narrow if all we know is win-lose.

The scientific field that deals with such matters is called game theory, used in military tactics and strategy, trade policy, corporate competition, limiting environmental pollution, and plans for nuclear war. The paradigmatic game is the Prisoner's Dilemma. It is very much non-zero-sum. Win-win, win-lose, and lose-lose outcomes are all possible. "Sacred" books carry few useful insights into strategy here. It is a wholly pragmatic game.

Imagine that you and a friend are arrested for committing a serious crime. For the purpose of the game, it doesn't matter whether either, neither, or both of you did it. What matters is that the police say they think you did. Before the two of you have any chance to compare stories or plan strategy, you are taken to separate interrogation cells. There, oblivious of your Miranda rights ("You have the right to remain silent . . ."), they try to make you confess. They tell you, as police sometimes do, that your friend has confessed and implicated you. (Some friend!) The police might be telling the truth. Or they might be lying. You're permitted only to plead innocent or guilty. If you're willing to say anything, what's your best tack to minimize punishment?

Here are the possible outcomes:

If you deny committing the crime and (unknown to you) your friend also denies it, the case might be hard to prove. In the plea bargain, both your sentences will be very light.

If you confess and your friend does likewise, then the effort the State had to expend to solve the crime was small. In exchange you both may be given a fairly light sentence, although not as light as if you both had asserted your innocence.

But if you plead innocent and your friend confesses, the state will ask for the maximum sentence for you and minimal punishment (maybe none) for your friend. Uh-oh. You're very vulnerable to a kind of double cross, what game theorists call "defection." So's he.

So if you and your friend "cooperate" with one another — both pleading innocent (or both pleading guilty) — you both escape the worst. Should you play it safe and guarantee a middle range of punishment by confessing? Then, if your friend pleads innocent while you plead guilty, well, too bad for him, and you might get off scot-free.

When you think it through, you realize that whatever your friend does you're better off defecting than cooperating. Maddeningly, the same holds true for your friend. But if you both defect, you're both worse off than if you had both cooperated. This is the Prisoner's Dilemma.

Now consider a repeated Prisoner's Dilemma, in which the two players go through a sequence of such games. At the end of each they figure out from their punishment how the other must have pled. They gain experience about each other's strategy (and character). Will they learn to cooperate game after game, both always denying that they committed any crime? Even if the reward for finking on the other is large?

You might try cooperating or defecting, depending on how the previous game or games have gone. If you cooperate overmuch, the other player may exploit your good nature. If you defect overmuch, your friend is likely to defect often, and this is bad for both of you. You know your defection pattern is data being fed to the other player. What is the right mix of cooperation and defection? How to behave then becomes, like any other question in Nature, a subject to be investigated experimentally.

This matter has been explored in a continuing round-robin computer tournament by the University of Michigan sociologist

Robert Axelrod in his remarkable book *The Evolution of Cooperation.* Various codes of behavior confront one another and at the end we see who wins (who gets the lightest cumulative prison term). The simplest strategy might be to cooperate all the time, no matter how much advantage is taken of you, or never to cooperate, no matter what benefits might accrue from cooperation. These are the Golden Rule and the Iron Rule. They always lose, the one from a superfluity of kindness, the other from an overabundance of ruthlessness. Strategies slow to punish defection lose — in part because they send a signal that noncooperation can win. The Golden Rule is not only an unsuccessful strategy; it is also dangerous for other players, who may succeed in the short term only to be mowed down by exploiters in the long term.

Should you defect at first, but if your opponent cooperates even once, cooperate in all future games? Should you cooperate at first, but if your opponent defects even once, defect in all future games? These strategies also lose. Unlike sports, you cannot rely on your opponent to be always out to get you.

The most effective strategy in many such tournaments is called "Tit-for-Tat." It's very simple: You start out cooperating, and in each subsequent round simply do what your opponent did last time. You punish defections, but once your partner cooperates, you're willing to let bygones be bygones. At first, it seems to garner only mediocre success. But as time goes on the other strategies defeat themselves, from too much kindness or cruelty, and this middle way pulls ahead. Except for always being nice on the first move, Tit-for-Tat is identical to the Brazen Rule. It promptly (in the very next game) rewards cooperation and punishes defection, and has the great virtue that it makes your strategy absolutely clear to your opponent. (Strategic ambiguity can be lethal.)

Once there get to be several players employing Tit-for-Tat, they rise in the standings together. To succeed, Tit-for-Tat strategists must find others who are willing to reciprocate, with whom they can cooperate. After the first tournament in which the Brazen Rule unexpectedly won, some experts thought the strategy too forgiving. Next tournament, they tried to exploit it by defecting more often. They always lost. Even experienced strategists tended to underestimate the power of forgiveness and reconciliation. Tit-for-Tat involves an inter-

TABLE OF PROPOSED RULES TO LIVE BY

The Golden Rule	Do unto others as you would have them do unto you.
The Silver Rule	Do not do unto others what you would not have them do unto you.
The Brazen (Brass) Rule	Do unto others as they do unto you.
The Iron Rule	Do unto others as you like, before they do it unto you.
The Tit-for-Tat Rule	Cooperate with others first, then do unto them as they do unto you.

esting mix of proclivities: initial friendliness, willingness to forgive, and fearless retaliation. The superiority of the Tit-for-Tat Rule in such tournaments has been recounted by Axelrod.

Something like it can be found throughout the animal kingdom and has been well-studied in our closest relatives, the chimps. Described and named "reciprocal altruism" by the biologist Robert Trivers, animals may do favors for others in expectation of having the favors returned — not every time, but often enough to be useful. This is hardly an invariable moral strategy, but it is not uncommon either. So there is no need to debate the antiquity of the Golden, Silver, and Brazen Rules, or Tit-for-Tat, and the priority of the moral prescriptives in the Book of Leviticus. Ethical rules of this sort were not originally invented by some enlightened human lawgiver. They go deep into our evolutionary past. They were with our ancestral line from a time before we were human.

The Prisoner's Dilemma is a very simple game. Real life is considerably more complex. If he gives our apple to the pencil man, is my father more likely to get an apple back? Not from the pencil man; we'll never see him again. But might widespread acts of charity improve the economy and give my father a raise? Or do we give the apple for emotional, not economic rewards? Also, unlike the players in an ideal Prisoner's Dilemma game, human beings and nations come to their interactions with predispositions, both hereditary and cultural.

But the central lessons in a not very prolonged round-robin of Prisoner's Dilemma are about strategic clarity; about the self-defeating nature of envy; about the importance of long-term over short-term goals; about the dangers of both tyranny and patsydom; and especially about approaching the whole issue of rules to live by as an experimental question. Game theory also suggests that a broad knowledge of history is a key survival tool.

Chapter 17

GETTYSBURG AND NOW*

This speech was delivered on July 3, 1988 to approximately 30,000 people at the 125th celebration of the Battle of Gettysburg and the rededication of the Eternal Light Peace Memorial, Gettysburg National Military Park, Gettysburg, Pennsylvania. Every quarter century, the peace memorial at Gettysburg is rededicated; Presidents Wilson, Franklin Roosevelt and Eisenhower have been the previous speakers.

From *Lend Me Your Ears: Great Speeches in History*, selected and introduced by William Safire (New York: W. W. Norton, 1992)

*Cowritten with Ann Druyan. The speech has been revised and updated for this book.

Fifty-one thousand human beings were killed or wounded here — ancestors of some of us, brothers of us all. This was the first full-fledged example of an industrialized war, with machine-tooled arms and railroad transport of men and matériel. It was the first hint of an age yet to come, our age; an intimation of what technology bent to the purposes of war might be capable. The new Spencer repeating rifle was used here. In May 1863, a reconnaissance balloon of the Army of the Potomac detected movement of Confederate troops across the Rappahannock River, the beginning of the campaign that led to the Battle of Gettysburg. That balloon was a precursor of air forces and strategic bombing and reconnaissance satellites.

A few hundred artillery pieces were deployed in the three-day battle of Gettysburg. What could they do? What was war like then? Here is an eyewitness account by Frank Haskel of Wisconsin, who fought on this battlefield for the Union Armies, of nightmarish, apparently hovering, cannon balls. It is from a letter to his brother:

> We could not often see the shell before it burst, but sometimes, as we faced towards the enemy and looked

> above our heads, the approach would be heralded by a prolonged hiss, which always seemed to me to be a line of something tangible terminating in a black globe distinct to the eye as the sound had been to the ear. The shell would seem to stop and hang suspended in the air an instant and then vanish in fire and smoke and noise. . . . Not ten yards away from us a shell burst among some bushes where sat three or four orderlies holding horses. Two of the men and one horse were killed.

It was a typical event from the Battle of Gettysburg. Something like this was repeated thousands of times. Those ballistic projectiles, launched from the cannons that you can see all over this Gettysburg Memorial, had a range, at best, of a few miles. The amount of explosive in the most formidable of them was some 20 pounds — roughly one-hundredth of a ton of TNT. It was enough to kill a few people.

But the most powerful chemical explosives used 80 years later, in World War II, were the blockbusters, so-called because they could destroy a city block. Dropped from aircraft, after a journey of hundreds of miles, each carried about 10 tons of TNT, a thousand times more than the most powerful weapon at the Battle of Gettysburg. A blockbuster could kill a few dozen people.

At the very end of World War II, the United States used the first atomic bombs to annihilate two Japanese cities. Each of those weapons, delivered after a voyage of sometimes a thousand miles, had the equivalent power of about 10,000 tons of TNT, enough to kill a few hundred thousand people. One bomb.

A few years later the United States and the Soviet Union developed the first thermonuclear weapons, the first hydrogen bombs. Some of them had an explosive yield equivalent to ten million tons of TNT; enough to kill a few million people. One bomb. Strategic nuclear weapons can now be launched to any place on the planet. Everywhere on Earth is a potential battlefield now.

Each of these technological triumphs advanced the art of mass murder by a factor of a thousand. From Gettysburg to the blockbuster, a thousand times more explosive energy; from the blockbuster to the atomic bomb, a thousand times more; and from the atomic bomb to the hydrogen bomb, a thousand times still more. A thousand times a thousand times a thousand is a billion; in less than one century, our most fearful weapon has become a billion times more deadly. But we have not become a billion times wiser in the generations that stretch from Gettysburg to us.

The souls that perished here would find the carnage of which we are now capable unspeakable. Today, the United States and the Soviet Union have booby-trapped our planet with almost 60,000 nuclear weapons. Sixty thousand nuclear weapons! Even a small fraction of the strategic arsenals could without question annihilate the two contending superpowers, probably destroy the global civilization, and possibly render the human species extinct. No nation, no man should have such power. We distribute these instruments of apocalypse all over our fragile world, and justify it on the grounds that it has made us safe. We have made a fool's bargain.

The 51,000 casualties here at Gettysburg represented one-third of the Confederate Army, and one-quarter of the Union Army. All those who died, with one or two exceptions, were soldiers. The best-known exception was a civilian in her own house who thought to bake a loaf of bread and, through two closed doors, was shot to death; her name was Jennie Wade. But in a global thermonuclear war, almost all the casualties would be civilians — men, women, and children, including vast numbers of citizens of nations that had no part in the quarrel that led to the war, nations far removed from the northern midlatitude "target zone." There would be billions of Jennie Wades. Everyone on Earth is now at risk.

In Washington there is a memorial to the Americans who died in the most recent major U.S. war, the one in Southeast Asia. Some 58,000 Americans perished, not a very different number from the casualties here at Gettysburg. (I ignore, as we too often do, the one or two million Vietnamese, Laotians, and Kampucheans who also died in that war.) Think of that dark, somber, beautiful, moving, touching memorial. Think of how long it is; actually, not much longer than a suburban street. 58,000 names. Imagine now that we are so foolish or inattentive as to permit a nuclear war to occur, and that, somehow, a similar memorial wall is built. How long would it have to be to contain the names of all those who will die in a major nuclear war? About a thousand miles long. It would stretch from here in Pennsylvania to Missouri. But, of course, there would be no one to build it, and few to read the roster of the fallen.

In 1945, at the close of World War II, the United States and the Soviet Union were virtually invulnerable. The United States — bounded east and west by vast and impassable oceans, north and south by weak and friendly neighbors — had the most effective armed forces, and the most powerful economy on the planet. We had nothing to fear. So we built nuclear weapons and their delivery systems. We initiated and vigorously pumped up an arms race with the Soviet Union. When we were done, all the citizens of the United States had handed their lives over to the leaders of the Soviet Union. Even today, post-Cold War, post-Soviet Union, if Moscow decides we should die, twenty minutes later we're dead. In nearly perfect symmetry, the Soviet Union had the largest standing army in the world in 1945, and no significant military threats to worry about. It joined the United States in the nuclear arms race so that today everyone in Russia has handed their lives over to the leaders of the United States. If Washington decides they should die, twenty minutes later they're dead. The lives of every American and every Russian citizen are now in the hands of a foreign power. I say we have made a fool's bargain. We — we Americans, we Russians — have spent 43 years and vast national treasure in making ourselves exquisitely vulnerable to instant annihilation. We have done it in the name of patriotism and "national security," so no one is supposed to question it.

Two months before Gettysburg, on May 3, 1863, there was a Confederate triumph, the battle of Chancellorsville. On the moonlit evening following the victory, General Stonewall Jackson and his staff, returning to the Confederate lines, were mistaken for Union cavalry. Jackson was shot twice in error by his own men. He died of his wounds.

We make mistakes. We kill our own.

There are some who claim that since we have not yet had an accidental nuclear war, the precautions being taken to prevent one must be adequate. But not three years ago we witnessed the disasters of the *Challenger* space shuttle and the Chernobyl nuclear power plant — high technology systems, one American, one Soviet, into which enormous quantities of national prestige had been invested. There were compelling reasons to prevent these disasters. In the preceding year, confident assertions were made by officials of both nations that no accidents of that sort could happen. We were not to worry. The experts would not permit an accident to happen. We have since learned that such assurances do not amount to much.

We make mistakes. We kill our own.

This is the century of Hitler and Stalin, evidence — if any were needed — that madmen can seize the reins of power of modern industrial states. If we are content in a world with nearly 60,000 nuclear weapons, we are betting our lives on the proposition that no present or future leaders, military or civilian — of the United States, the Soviet Union, Britain, France, China, Israel, India, Pakistan, South Africa, and whatever other nuclear powers there will be — will ever stray from the strictest standards of prudence. We are gambling on their sanity and sobriety even in times of great personal and national crisis — all of them, for all times to come. I say this is asking too much of us. Because we make mistakes. We kill our own.

The nuclear arms race and the attendant Cold War cost something. They don't come free. Apart from the immense diversion of fiscal and intellectual resources away from the civilian economy, apart from the psychic cost of living out our lives under the Damoclean sword, what has been the price of the Cold War?

Between the beginning of the Cold War in 1946, and its end in 1989, the United States spent (in equivalent 1989 dollars) well over $10 trillion in its global confrontation with the Soviet Union. Of this sum, more than a third was spent by the Reagan Administration, which added more to the national debt than all previous administrations, back to George Washington, combined. At the beginning of the Cold War, the nation was, in all significant respects, untouchable by any foreign military force. Today, after the expenditure of this immense national treasure (and despite the end of the Cold War), the United States is vulnerable to virtually instant annihilation.

A business that spent its capital so recklessly, and with so little effect, would have been bankrupt long ago. Executives who could not recognize so clear a failure of corporate policy would long before now have been dismissed by the stockholders.

What else could the United States have done with that money (not all of it, because prudent defense is, of course, necessary — but, say, half of it)? For a little over $5 trillion, skillfully applied, we could have made major progress toward eliminating hunger, homelessness, infectious disease, illiteracy, ignorance, poverty, and safeguarding the environment — not just in the United States but worldwide. We could have helped make the planet agriculturally self-sufficient and removed many of the causes of violence and war. And this could have been done with enormous benefit to the American economy. We could have made deep inroads into the national debt. For less than a percent of that money, we could have mustered a long-term international program of manned exploration of Mars. Prodigies of human inventiveness in art, architecture, medicine, and science could be supported for decades with a tiny fraction of that money. The technological and entrepreneurial opportunities would have been prodigious.

Have we been wise in spending so much of our vast wealth on the preparations and paraphernalia of war? At the present time we are still spending at Cold War levels. We have made a fool's bargain. We have been locked in a deadly embrace with the Soviet Union, each side always propelled by the abundant malefactions of the other; almost always looking to the short term — to the next Congressional or Presidential election, to the next Party Congress — and almost never seeing the big picture.

Dwight Eisenhower, who was closely associated with this Gettysburg community, said, "The problem in defense spending is to figure out how far you should go without destroying from within what you are trying to defend from without." I say we have gone too far.

How do we get out of this mess? A Comprehensive Test Ban Treaty would stop all future nuclear weapons tests; they are the chief technological driver that propels, on both sides, the nuclear arms race. We need to abandon the ruinously expensive notion of Star Wars, which cannot protect the civilian population from nuclear war and subtracts from, not adds to, the national security of the United States. If we want to enhance deterrence, there are far better ways to do it. We need to make safe, massive, bilateral, intrusively inspected reductions in the strategic and tactical nuclear arsenals of the United States, Russia, and all other nations. (The INF and START treaties represent tiny steps, but in the right direction.) That's what we should be doing.

Because nuclear weapons are comparatively cheap, the big ticket item has always been, and remains, conventional military forces. An extraordinary opportunity is now before us. The Russians and the Americans have been engaged in major conventional force reductions in Europe. These should be extended to Japan, Korea, and other nations perfectly well able to defend themselves. Such conventional force reduction is in the interest of peace, and in the interest of a sane and healthy American economy. We ought to meet the Russians halfway.

The world today spends $1 trillion a year on military preparations, most of it on conventional arms. The United States and Russia are the leading arms merchants. Much of that money is spent only because the nations of the world are unable to take the unbearable step of reconciliation with their adversaries (and some of it because governments need forces to suppress and intimidate their own people). That trillion dollars a year takes food from the mouths of poor people. It cripples potentially effective economies. It is a scandalous waste, and we should not countenance it.

It is time to learn from those who fell here. And it is time to act.

In part the American Civil War was about freedom; about extending the benefits of the American Revolution to all Americans, to make valid for everyone that tragically unfulfilled promise of "liberty and justice for all." I'm concerned about a lack of historical pattern recognition. Today the fighters for freedom do not wear three-cornered hats and play the fife and drum. They come in other costumes. They may speak other languages. They may adhere to other religions. The color of their skin may be different. But the creed of liberty means nothing if it is only our own liberty that excites us. People elsewhere are crying, "No taxation without representation," and in Western and Eastern Africa, or the West Bank of the Jordan River, or Eastern Europe, or Central America they are shouting in increasing numbers, "Give us liberty or give us death." Why are we unable to hear most of them? We Americans have powerful nonviolent means of persuasion available to us. Why are we not using these means?

The Civil War was mainly about union; union in the face of differences. A million years ago, there were no nations on the planet. There were no tribes. The humans who were here were divided into small family groups of a few dozen people each. We wandered. That was the horizon of our identification, an itinerant family group. Since then, the horizons have expanded. From a handful of hunter-gatherers, to a tribe, to a horde, to a small city-state, to a nation, and today to immense nation-states. The average person on the Earth today owes his or her primary allegiance to a group of something like 100 million people. It seems very clear that if we do not destroy ourselves first, the unit of primary identification of most human beings will before long be the planet Earth and the human species. To my mind, this raises the key question: whether the fundamental unit of identification will expand to embrace the planet and the species, or whether we will destroy ourselves first. I'm afraid it's going to be very close.

The identification horizons were broadened in this place 125 years ago, and at great cost to North and South, to blacks and whites. But we recognize that expansion of identification horizons as just. Today there is an urgent, practical necessity to work together on arms control, on the world economy, on the global environment. It is clear that the nations of the world now can only rise and fall together. It is not a question of one nation winning at the expense of another. We must all help one another or all perish together.

On occasions like this it is customary to quote homilies — phrases by great men and women that we've all heard before. We hear, but we tend not to focus. Let me mention one, a phrase that was uttered not far from this spot by Abraham Lincoln: "With malice toward none, with charity for all . . ." *Think* of what that means. This is what is expected of us, not merely because our ethics command it, or because our religions preach it, but because it is necessary for human survival.

Here's another: "A house divided against itself cannot stand." Let me vary it a little: A species divided against itself cannot stand. A planet divided against itself cannot stand. And to be inscribed on this Eternal Light Peace Memorial, which is about to be rekindled and rededicated, is a stirring phrase: "A World United in the Search for Peace."

The real triumph of Gettysburg was not, I think, in 1863 but in 1913, when the surviving veterans, the remnants of the adversary forces, the Blue and the Gray, met in celebration and solemn memorial. It had been the war that set brother against brother, and when the time came to remember, on the 50th anniversary of the battle, the survivors fell, sobbing, into one another's arms. They could not help themselves.

It is time now for us to emulate them — NATO and the Warsaw Pact, Tamils and Singhalese, Israelis and Palestinians, whites and blacks, Tutsis and Hutus, Americans and Chinese, Bosnians and Serbs, Unionists and Ulsterites, the developed and the underdeveloped worlds.

We need more than anniversary sentimentalism and holiday piety and patriotism. Where necessary, we must confront and challenge the conventional wisdom. It is time to learn from those who fell here. Our challenge is to reconcile, not *after* the carnage and the mass murder, but *instead* of the carnage and the mass murder. It is time to fly into one another's arms.

It is time to act.

Update: To some degree, we have. In the time since this address was delivered, we Americans, we Russians, we humans have made major reductions in our nuclear arsenals and delivery systems — but not nearly enough for safety. We seem to be on the verge of a Comprehensive Test Ban Treaty — but the means of assembling and conveying nuclear warheads has spread or is about to spread to many more nations.

This circumstance is often described as the exchange of one potential catastrophe for another, with no substantial improvement. But a handful of nuclear weapons, as catastrophic as they can be — as much human tragedy as they could cause — are as toys compared with the 60 or 70 thousand nuclear weapons that the United States and the Soviet Union accumulated at the height of the Cold War. Sixty or seventy thousand nuclear weapons could destroy the global civilization and possibly even the human species. The arsenals that North Korea or Iraq or Libya or India or Pakistan could accumulate cannot, in the foreseeable future, do any of that.

At the other extreme is the boast by American political leaders that no Russian nuclear weapon is targeted on a U.S. child or city. This may well be right, but retargeting takes at most 15 or 20 minutes. And both the United States and Russia retain thousands of nuclear weapons and delivery systems. That is why, throughout this book, I have insisted that nuclear weapons remain our greatest danger — even though substantial, even stunning, improvements in human safety have transpired. But it could all change overnight.

In Paris, in January 1993, 130 nations signed the Chemical Weapons Convention. After more than 20 years of negotiation, the world declared its readiness to outlaw these weapons of mass destruction. But as I write these words, the United States and Russia have still not ratified the Convention. What are we waiting for? Meanwhile, Russia has not yet ratified the START II accords, which would reduce the American and Russian strategic nuclear arsenals by 50 percent, down to 3,500 deployed warheads each.

Since the end of the Cold War, the American military budget has declined — but only by 10 or 15 percent, and almost none of that saving seems to have been effectively applied to the civilian economy. The Soviet Union has collapsed — but widespread misery and instability in the region is reason for worry about the global future. Democracy has to some extent reasserted itself in Eastern Europe, and Central and South America — but has made few inroads, except for Taiwan and South Korea, in East Asia; and is distorted in Eastern Europe by the worst excesses of capitalism. Identification horizons have broadened in Western Europe — but generally narrowed in the United States and the former Soviet Union. Progress has been made at reconciliation in Northern Ireland and in Israel/Palestine — but terrorists are still able to hold the peace process hostage.

Draconian cuts in the U.S. federal budget must be made, we are told, because of an urgent need to balance the budget. But, oddly, an institution whose share of the gross domestic product is higher than the entire federal discretionary budget is essentially off-limits. This is the military's $264 billion (compared with $17 billion for all civilian science and space programs). Actually, if hidden military costs and the intelligence budget were included, the military's share would be much larger.

With the Soviet Union vanquished, what is this immense sum of money for? Russia's annual military budget is about $30 billion. So is China's. The combined military budgets of Iran, Iraq, North Korea, Syria, Libya, and Cuba amount to about $27 billion. The U.S. is outspending all of them put together by a factor of three. It accounts for about 40 percent of world military expenditures.

The Clinton defense budget for fiscal year 1995 was some $30 billion higher than Richard Nixon's defense budget in the height of the Cold War, 20 years earlier. With the Republican-proposed increments, the U.S. defense budget will grow in real dollars by 50 percent by the year 2000. There is no effective voice in either political party opposing such growth — even as agonizing rips in the social safety net are being planned.

Our skinflint Congress turns shockingly profligate when it comes to the military, pressing unsolicited billions on a Department of Defense trying to exercise some modicum of self-control. While freighters in busy harbors and embassy pouches immune to border inspection are now the most likely delivery systems for nuclear weapons to American soil, there is strong Congressional pressure for space-based interceptors to protect the United States from the nonexistent intercontinental ballistic missiles of rogue nations. Wacky $2.3 billion rebate schemes are proposed for foreign nations so they can buy American arms. Taxpayers' money is given to American aerospace companies so they can buy other American aerospace companies. Around $100 billion is spent every year to defend Western Europe, Japan, South Korea, and other nations — virtually all of which enjoy healthier balances of trade than the United States does. We plan to keep almost 100,000 troops stationed in Western Europe indefinitely. To defend against whom?

Meanwhile, the hundreds of billions of dollars it will cost to clean up the military nuclear and chemical waste are a burden passed on to our children that we somehow don't have much problem with. Why do we have such difficulty grasping that national security is a much deeper and more subtle matter than the number of rocks in our pile? Despite all the talk of the military budget's being "cut to the bone," in the world we live in, it's bulging with marbleized fat. Why should the military budget be sacrosanct when so much else that our national well-being depends on is in danger of being thoughtlessly destroyed?

There is much left to do. It is still time to act.

Chapter 18

THE TWENTIETH CENTURY

To realize in its completeness the universal beauty and perfection of the works of God, we must recognize a certain perpetual and very free progress of the whole universe . . . [T]here always remain in the abyss of things slumbering parts which have yet to be awakened . . .

GOTTFRIED WILHELM LEIBNIZ,
On the Ultimate Origination of Things
(1697)

Society never advances. It recedes as fast on one side as it gains on the other. It undergoes continual changes; it is barbarous, it is civilized, it is christianized, it is rich, it is scientific; but . . . for everything that is given something is taken.

RALPH WALDO EMERSON,
"Self-Reliance," *Essays: First Series* (1841)

The twentieth century will be remembered for three broad innovations: unprecedented means to save, prolong, and enhance life; unprecedented means to destroy life, including for the first time putting our global civilization at risk; and unprecedented insights into the nature of ourselves and the Universe. All three of these developments have been brought forth by science and technology, a sword with two razor-sharp edges. All three have roots in the distant past.

SAVING, PROLONGING, AND ENHANCING HUMAN LIFE

Until about ten thousand years ago, with the invention of agriculture and the domestication of animals, the human food supply was limited to fruits and vegetables in the natural environment and game animals. But the sparsity of naturally grown foodstuffs was such that the Earth could maintain no more than about 10 million or so of us. In contrast, by the end of the twentieth century, there will be six billion people. That means that 99.9 percent of us owe our lives to agricultural technology and the science that underlies it — plant and animal genetics and behavior, chemical fertilizers, pesticides, preservatives, plows, combines and other agricultural implements, irrigation — and refrigeration in trucks, railway cars, stores, and homes. Many of the most striking advances in agricultural technology — including the "Green Revolution" — are products of the twentieth century.

Through urban and rural sanitation, clean water, other public health measures, acceptance of the germ theory of disease, antibiotics and other pharmaceuticals, and genetics and molecular biology, medical science has enormously improved the well-being of people all over the world — but especially in the developed countries. Smallpox has been eradicated worldwide, the area of the Earth in which malaria flourishes shrinks year by year, and diseases I remember from my childhood, such as whooping cough, scarlet fever, and polio, are almost gone today. Among the most important twentieth-century inventions are comparatively inexpensive birth control methods — which, for the first time, permit women safely to control their reproductive destinies, and are working the emancipation of half of the human species. They permit major declines in the perilously increasing populations in many countries without requiring oppressive restrictions on sexual activity. It is also true that the chemicals and radiation produced by our technology have induced new diseases and are implicated in cancer. The global proliferation of cigarettes leads to an estimated 3 million deaths a year (all of course preventable). By 2020, the number is estimated, by the World Health Organization, to reach 10 million a year.

But technology has given much more than it has taken away. The clearest sign of this is that life expectancy in the United States and Western Europe in 1901 was about 45 years, while today it is approaching 80 years, a little more for women, a little less for men. Life expectancy is probably the single most effective index of quality of life: If you're dead, you're probably not having a good time. That said, there are still a billion of us without enough to eat, and 40,000 children dying needlessly every day on our planet.

Through radio, television, phonographs, audiotape players, compact discs, telephones, fax machines, and computer information networks, technology has profoundly changed the face of popular culture. It has made possible the pros and cons of global entertainment, of multinational corporations with loyalties to no particular country, transnational affinity groups, and direct access to the political and religious views of other cultures. As we saw in the highly attenuated rebellion at Tiananmen Square and the one at the "White House" in Moscow, faxes, telephones, and computer networks can be powerful tools of political upheaval.

The introduction of mass-market paperback books in the 1940s has brought the world's literature and the insights of its greatest thinkers, present and past, into the lives of ordinary people. And even if the price of paperback books is soaring today, there are still great bargains, such as the dollar-a-volume classics from Dover Books. Along with progress in literacy such trends are the allies of Jeffersonian democracy. On the other hand what passes for literacy in America in the late twentieth century is a very rudimentary knowledge of the English language, and television in particular tends to seduce the mass population away from reading. In pursuit of the profit motive, it has dumbed itself down to lowest-common-denominator programming — instead of rising up to teach and inspire.

From paper clips, rubber bands, hair dryers, ballpoint pens, computers, dictating and copying machines, electric mixers, microwave ovens, vacuum cleaners, dish and clothes washers and driers, widespread interior and street lights, to automobiles, aviation, machine tools, hydroelectric power plants, assembly line manufacturing, and enormous construction equipment, the technology of our century has eliminated drudgery, created more leisure time, and enhanced the lives of many. It has also upended many of the routines and conventions that were prevalent in 1901.

The use of potentially life-saving technology differs from nation to nation. The United States, for example, has the highest infant mortality of any industrial nation. It has more young black men in prison than in college, and a greater percentage of its citizens in jail than any other industrial nation. Its students routinely perform poorly on standardized science and mathematics tests when compared with students of the same age in other countries. The disparity in real income between the rich and the poor and the decline of the middle class have been growing swiftly over the last decade and a half. The United States is last among industrialized nations in the fraction of the national income given each year to help people in other countries. High technology industry has been fleeing American shores. After leading the world in almost all these respects in midcentury, there are some signs of decline in the United States at century's end. The quality of leadership can be pointed to, but so can the dwindling penchant for critical thinking and political action in its citizens.

TOTALITARIAN AND MILITARY TECHNOLOGY

The means of making war, of mass killings, of the annihilations of whole peoples, has reached unprecedented levels in the twentieth century. In 1901, there were no military aircraft or missiles, and the most powerful artillery could loft a shell a few miles and kill a handful of people. By the second third of the twentieth century, some 70,000 nuclear weapons had been accumulated. Many of them were fitted to strategic rocket boosters, fired from silos or submarines, able to reach virtually any part of the world, and each warhead powerful enough to destroy a large city. Today we are in the throes of major arms reductions, both in warheads and delivery systems, by the United States and the former Soviet Union, but we will be able to annihilate the global civilization into the foreseeable future. In addition, horrendously deadly chemical and biological weapons are in many hands worldwide. In a century bubbling over with fanaticism, ideological certainty, and mad leaders, this accumulation of unprecedentedly lethal weapons does not bode well for the human future. Over 150 million human beings have been killed in warfare and by the direct orders of national leaders in the twentieth century.

Our technology has become so powerful that not only on purpose but inadvertently we have become able to alter the environment on a large scale, and threaten many species on Earth, our own included. The simple fact is that we are performing unprecedented experiments on the global environment and in general hoping against hope that the problems will solve themselves and go away. The one bright spot is the Montreal Protocol and ancillary international agreements by which the industrial nations of the world agreed to phase out production of CFCs and other chemicals that attack the ozone layer. But in reducing carbon dioxide emissions to the atmosphere, in solving the problem of chemical and radioactive wastes, and in other areas progress has been slow to dismal.

Ethnocentric and xenophobic vendettas have been rife on every continent. Systematic attempts to annihilate whole ethnic groups have occurred — most notably in Nazi Germany, but also in Rwanda, the former Yugoslavia, and elsewhere. Similar tendencies have existed throughout human history, but only in the twentieth century has technology made killing on such a scale practical. Strategic bombing, missiles, and long-range artillery have the "advantage" that the combatants need not come face-to-face with the agony they have worked. Their consciences need not trouble them. The global military budget at the end of the twentieth century is close to a trillion dollars a year. Think of how much human good could be purchased for even a fraction of that sum.

The twentieth century has been marked by the collapse of monarchies and empires and the rise of at least nominal democracies — as well as many ideological and military dictatorships. The Nazis had a list of reviled groups they set out to systematically exterminate: Jews, gays and lesbians, socialists and communists, the handicapped, and people of African origin (of whom there were almost none in Germany). In the militantly "pro-life" Nazi regime, women were relegated to *"Kinder, Küche, Kircher"* — children, cooking, the church.* How affronted a good Nazi would be at an American society that, more than any other country, dominates the planet, in which Jews, homosexuals, the handicapped, and people of African origin have full legal rights, socialists are at least tolerated in principle, and women are entering the workplace in record numbers. But only around 11 percent of the members of the U.S. House of Representatives are women, instead of a little more than 50 percent, as it would be if proportional representation were practiced. (The corresponding number for Japan is 2 percent.)

Thomas Jefferson taught that a democracy was impractical unless the people were educated. No matter how stringent the protections of the people might be in constitutions or common law, there would

*After outlining traditional Christian views of women from patristic times to the Reformation, the Australian philosopher John Passmore (*Man's Responsibility for Nature: Ecological Problems and Western Traditions* [New York: Scribner's, 1974]) concludes that *Kinder, Küche, Kircher* "as a description of the role of women is not an invention of Hitler's, but a typical Christian slogan."

always be a temptation, Jefferson thought, for the powerful, the wealthy, and the unscrupulous to undermine the ideal of government run by and for ordinary citizens. The antidote is vigorous support for the expression of unpopular views, widespread literacy, substantive debate, a common familiarity with critical thinking, and skepticism of pronouncements of those in authority — which are all also central to the scientific method.

THE REVELATIONS OF SCIENCE

Every branch of science has made stunning advances in the twentieth century. The very foundations of physics have been revolutionized by the special and general theories of relativity, and by quantum mechanics. It was in this century that the nature of atoms — with protons and neutrons in a central nucleus and electrons in a surrounding cloud — was first understood, when the constituent components of protons and neutrons, the quarks, were first glimpsed, and when a host of exotic short-lived elementary particles first showed up under the ministrations of high energy accelerators and cosmic rays. Fission and fusion have made possible the corresponding nuclear weapons, fission power plants (a not-unmixed blessing), and the prospect of fusion power plants. An understanding of radioactive decay has given us definitive knowledge of the age of the Earth (about 4.6 billion years) and of the time of the origin of life on our planet (roughly 4 billion years ago).

In geophysics, plate tectonics was discovered — a set of conveyer belts under the Earth's surface carrying continents from birth to death, and moving at a rate of about an inch a year. Plate tectonics is essential for understanding the nature and history of landforms and the topography of the sea bottoms. A new field of planetary geology has emerged in which the landforms and interior of the Earth can be compared with those of other planets and their moons, and the chemistry of rocks on other worlds — determined either remotely or from returned samples brought back by spacecraft or from meteorites now recognized as arising from other worlds — can be compared with the rocks on Earth. Seismology has plumbed the structure of the deep interior of the Earth and discovered beneath the crust a semi-liquid mantle, a liquid iron core, and a solid inner core — all of which must be explained if we wish to know the processes by which our planet came to be. Some mass extinctions of life in the past are now understood by immense mantle plumes gushing up through the surface and generating lava seas where solid land once stood. Others are due to the impact of large comets or near-Earth asteroids igniting the skies and changing the climate. In the next century, at the very least we ought to be inventorying comets and asteroids to see if any of them has our name on it.

One cause for scientific celebration in the twentieth century is the discovery of the nature and function of DNA, deoxyribonucleic acid — the key molecule responsible for heredity in humans and in most other plants and animals. We have learned to read the genetic code and in increasing numbers of organisms we have mapped all the genes and know what functions of the organisms most of them are in charge of. Geneticists are well on their way to mapping the human genome — an accomplishment with enormous potential for both good and evil. The most significant aspect of the DNA story is that the fundamental processes of life now seem fully understandable in terms of physics and chemistry. No life force, no spirit, no soul seems to be involved. Likewise in neurophysiology: Tentatively, the mind seems to be the expression of the hundred trillion neural connections in the brain, plus a few simple chemicals.

Molecular biology now permits us to compare any two species, gene by gene, molecular building block by molecular building block, to uncover the degree of relatedness. These experiments have shown conclusively the deep similarity of all beings on Earth and have confirmed the general relations previously found by evolutionary biology. For example, humans and chimpanzees share 99.6 percent of their active genes, confirming that chimps are our closest relatives, and that we share with them a recent common ancestor.

In the twentieth century for the first time field researchers have lived with other primates, carefully observing their behavior in their natural habitats, and discovering compassion, foresight, ethics, hunting, guerrilla warfare, politics, tool use, tool manufacture, music, rudimentary nationalism, and a host of other characteristics previously thought to be uniquely human. The debate on chimpanzee language abilities is still ongoing. But there is a bonobo (a "pigmy chimp") in Atlanta named Kanzi who easily uses a symbolic language of several hundred characters and who has also taught himself to manufacture stone tools.

Many of the most striking recent advances in chemistry are connected with biology, but let me mention one that is of much broader significance: the nature of the chemical bond has been understood, the forces in quantum physics that determine which atoms like to link up with which other atoms, how strongly, and in what configuration. It has also been found that radiation applied to not implausible primitive atmospheres for the Earth and other planets generate amino acids and other key building blocks of life. Nucleic acids and other molecules in the test tube have been found to reproduce themselves and reproduce their mutations. Thus substantial progress has been made in the twentieth century toward understanding and generating the origin of life. Much of biology is reducible to chemistry and much of chemistry is reducible to physics. This is not yet completely true, but the fact that it is even a little bit true is a most important insight into the nature of the Universe.

Physics and chemistry, coupled with the most powerful computers on Earth, have tried to understand the climate and general circulation of the Earth's atmosphere through time. This powerful tool is used to evaluate the future consequences of the continued emission of CO_2 and other greenhouse gases into the Earth's atmosphere. Meanwhile, much easier, meteorological satellites permit weather prediction at least days in advance, avoiding billions of dollars in crop failures every year.

At the beginning of the twentieth century astronomers were stuck at the bottom of an ocean of turbulent air and left to peer at distant worlds. By the end of the twentieth century great telescopes are in Earth orbit peering at the heavens in gamma rays, X rays, ultraviolet light, visible light, infrared light, and radio waves.

Marconi's first radio broadcast across the Atlantic Ocean occurred in 1901. We have now used radio to communicate with four spacecraft beyond the outermost known planet of our Solar System, and to hear the natural radio emission from quasars 8 and 10 billion light-years away — as well as the so-called black body background radiation, the radio remnants of the Big Bang, the vast explosion that began the current incarnation of the Universe.

Exploratory spacecraft have been launched to study 70 worlds and to land on three of them. The century has seen the almost mythic accomplishment of sending 12 humans to the Moon and bringing them, and over a hundred kilograms of moon rocks, back safely. Robotic craft have confirmed that Venus, driven by a massive greenhouse effect, has a surface temperature of almost 900° Fahrenheit; that 4 billion years ago Mars had an Earth-like climate; that organic molecules are falling from the sky of Saturn's moon Titan like manna from Heaven; that comets are made of perhaps a quarter of organic matter.

Four of our spacecraft are on their way to the stars. Other planets have recently been found around other stars. Our Sun is revealed to be in the remote outskirts of a vast, lens-shaped galaxy comprising some 400 billion other suns. At the beginning of the century it was thought that the Milky Way was the only galaxy. We now recognize that there are a hundred billion others, all fleeing one from another as if they are the remnants of an enormous explosion, the Big Bang. Exotic denizens of the cosmic zoo have been discovered that were not even dreamt of at the turn of the century — pulsars, quasars, black holes. Within observational reach may be the answers to some of the deepest questions humans have ever asked — on the origin, nature, and fate of the entire Universe.

Perhaps the most wrenching by-product of the scientific revolution has been to render untenable many of our most cherished and most comforting beliefs. The tidy anthropocentric proscenium of our ancestors has been replaced by a cold, immense, indifferent Universe in which humans are relegated to obscurity. But I see the emergence in our consciousness of a Universe of a magnificence, and an intricate, elegant order far beyond anything our ancestors imagined. And if much about the Universe can be understood in terms of a few simple laws of Nature, those wishing to believe in God can certainly ascribe those beautiful laws to a Reason underpinning all of Nature. My own view is that it is far better to understand the Universe as it really is than to pretend to a Universe as we might wish it to be.

Whether we will acquire the understanding and wisdom necessary to come to grips with the scientific revelations of the twentieth century will be the most profound challenge of the twenty-first.

Chapter 19

IN THE VALLEY OF THE SHADOW

Is this, then, true or mere vain fantasy?

EURIPIDES,
Ion (ca. 410 B.C.)

Six times now have I looked Death in the face. And six times Death has averted his gaze and let me pass. Eventually, of course, Death will claim me — as he does each of us. It's only a question of when. And how.

I've learned much from our confrontations — especially about the beauty and sweet poignancy of life, about the preciousness of friends and family, and about the transforming power of love. In fact, almost dying is such a positive, character-building experience that I'd recommend it to everybody — except, of course, for the irreducible and essential element of risk.

I would love to believe that when I die I will live again, that some thinking, feeling, remembering part of me will continue. But as much as I want to believe that, and despite the ancient and worldwide cultural traditions that assert an afterlife, I know of nothing to suggest that it is more than wishful thinking.

I want to grow really old with my wife, Annie, whom I dearly love. I want to see my younger children grow up and to play a role in their character and intellectual development. I want to meet still unconceived grandchildren. There are scientific problems whose outcomes I long to witness — such as the exploration of many of the worlds in our Solar System and the search for life elsewhere. I want to learn how major trends in human history, both hopeful and worrisome, work themselves out: the dangers and promise of our technology, say; the emancipation of women; the growing political, economic, and technological ascendancy of China; interstellar flight.

If there were life after death, I might, no matter when I die, satisfy most of these deep curiosities and longings. But if death is nothing more than an endless dreamless sleep, this is a forlorn hope. Maybe this perspective has given me a little extra motivation to stay alive.

The world is so exquisite, with so much

love and moral depth, that there is no reason to deceive ourselves with pretty stories for which there's little good evidence. Far better, it seems to me, in our vulnerability, is to look Death in the eye and to be grateful every day for the brief but magnificent opportunity that life provides.

For years, near my shaving mirror — so I see it every morning — I have kept a framed postcard. On the back is a penciled message to a Mr. James Day in Swansea Valley, Wales. It reads:

> *Dear Friend,*
>
> *Just a line to show that I am alive & kicking and going grand. It's a treat.*
>
> *Yours,*
> *WJR*

It's signed with the almost-indecipherable initials of one William John Rogers. On the front is a color photo of a sleek, four-funneled steamer captioned "White Star Liner *Titanic*." The postmark was imprinted the day before the great ship went down, losing more than 1,500 lives, including Mr. Rogers's. Annie and I display the postcard for a reason. We know that "going grand" can be the most temporary and illusory state. So it was with us.

We were in apparent good health, our children thriving. We were writing books, embarking on ambitious new television and motion picture projects, lecturing, and I continued to be engaged in the most exciting scientific research.

Standing by the framed postcard one morning late in 1994, Annie noticed an ugly black-and-blue mark on my arm that had been there for many weeks. "Why hasn't it gone away?" she asked. So at her insistence I somewhat reluctantly (black-and-blue marks can't be serious, can they?) went to the doctor to have some routine blood tests.

We heard from him a few days later when we were in Austin, Texas. He was troubled. There clearly was some lab mixup. The analysis showed the blood of a very sick person. "Please," he urged, "get retested right away." I did. There had been no mistake.

My red cells, which carry oxygen all over the body, and my white cells, which fight disease, were both severely depleted. The most likely explanation: that there was a problem with the stem cells, the common ancestors of both white and red blood cells, which are generated in the bone marrow. The diagnosis was confirmed by experts in the field. I had a disease I had never heard of before, myelodysplasia. Its origin is nearly unknown. If I did nothing, I was astonished to hear, my chances were zero. I'd be dead in six months. I was still feeling fine — perhaps a little lightheaded from time to time. I was active and productive. The notion that I was on death's doorstep seemed like a grotesque joke.

There was only one known means of treatment that might generate a cure: a bone marrow transplant. But that would work only if I could find a compatible donor. Even then, my immune system would have to be entirely suppressed so the donor's bone marrow wouldn't be rejected by my body. However, a severely suppressed immune system might kill me in several other ways — for example, by so limiting my resistance to disease that I might fall prey to any passing microbe. Briefly I thought about doing nothing and waiting for the advance in medical research to find a new cure. But that was the slimmest of hopes.

All our lines of research as to where to go converged on the Fred Hutchinson Cancer Research Center in Seattle, one of the premier institutions for bone marrow transplant in the world. It is where many experts in the field hang their hats — among them E. Donnall Thomas, the winner of the 1990 Nobel Prize in Physiology and Medicine, for perfecting the present techniques of bone marrow transplantation. The high competence of doctors and nurses, the excellence of the care, fully justified the advice we were given to be treated at "the Hutch."

The first step was to see if a compatible donor could be found. Some people never find one. Annie and I called my only sibling — my younger sister, Cari. I found myself allusive and indirect. Cari didn't even know I was ill. Before I could get to the point, she said, "You got it. Whatever it is . . . liver . . . lung. . . . It's yours." I still get a lump in my throat every time I think of Cari's generosity. But there was of course no guarantee that her bone marrow would be compatible with mine. She underwent a series of tests, and one after another, all six compatibility factors matched mine. She was a perfect match. I was incredibly lucky.

"Lucky" is a comparative term, though. Even with the perfect compatibility, my overall chances of a cure were something like 30 percent. That's like playing Russian roulette with four cartridges instead of one in the cylinder. But it was by far the best chance I had, and I had faced longer odds in the past.

Our whole family moved to Seattle, including Annie's parents. We enjoyed a constant flow of visitors — grown-up children, my grandson, other relatives and friends — both when I was in the hospital and when I was an outpatient. I'm sure that the support and love I received, especially from Annie, tilted the odds in my favor.

There were, as you might guess, many scary aspects. I remember one night, on medical instructions, getting up at 2 A.M. and opening the first of 12 plastic containers of busulfan tablets, a potent chemotherapeutic agent. The bag read:

CHEMOTHERAPY DRUG
BIOHAZARD BIOHAZARD
TOXIC
Dispose of as BIOHAZARD

One after another, I popped 72 of these pills. It was a lethal amount. If I was not to have a bone marrow transplant soon after, this immune suppression therapy by itself would have killed me. It was like taking a fatal dose of arsenic or cyanide, and hoping that the right antidote would be supplied in time.

The drugs to suppress my immune system had a few direct effects. I was in a continuous state of moderate nausea, but it was controlled by other drugs and not so bad that I couldn't get some work done. I lost almost all my hair — which, along with a later weight loss, gave me a somewhat cadaverous appearance. But I was much buoyed when our four-year-old son, Sam, looked me over and said: "Nice haircut, Dad." And then, "I don't know anything about you being sick. All I know is, you're gonna get better."

I had expected the transplant itself to be enormously painful. It was nothing of the sort. It was just like a blood transfusion, with my sister's bone marrow cells on their own finding their way to my own bone marrow. *Some* aspects of the treatment were extremely painful, but there's a kind of traumatic amnesia that happens, so that when it's all over you've almost forgotten the pain. The Hutch has an enlightened policy of self-administered antipain drugs, including morphine derivatives, so that I could immediately deal with severe pain. It made the whole experience much more bearable.

At the end of the treatment, my red and white cells were mainly Cari's. The sex chromosomes were XX, instead of the rest of my body's XY. I had female cells and platelets circulating through my body. I kept waiting for some of Cari's interests to assert themselves — a passion for riding horses, say, or for seeing half a dozen Broadway plays at one clip — but it never happened.

Annie and Cari saved my life. I'll always be grateful to them for their love and compassion. After being released from the hospital, I needed all sorts of medical attention, including drugs administered several times a day through a portal in my vena cava. Annie was my "designated caregiver" — administering medication day and night, changing dressings, checking vital signs, and providing essential support. People who arrive at the hospital alone are said, understandably, to have much poorer chances.

I was, for the moment, spared because of medical research. Some of it was applied research, designed to help cure or mitigate killer diseases directly. Some of it was basic research, designed only to understand how living things work — but with ultimate unpredictable practical benefits, serendipitous results.

I was spared also by the medical insurance provided by Cornell University and (as a spousal benefit via Annie) by the Writers Guild of America — the organization of writers for movies, television, etc. There are tens of millions in America without such medical insurance. What would we have done in their shoes?

In my writings, I have tried to show how closely related we are to other animals, how cruel it is to inflict pain on them, and how morally bankrupt it is to slaughter them to, say, manufacture lipstick. But still, as Dr. Thomas put it in his Nobel Prize lecture, "The marrow grafting could not have reached clinical application without animal research, first in in-bred rodents and then in out-bred species, particularly the dog." I remain very conflicted on this issue. I would not be alive today if not for research on animals.

So life returned to normal. Annie and I and our family returned to Ithaca, New York, where we live. I completed several research projects and did the final proofing of my book *The Demon-Haunted World: Science as a Candle in the Dark*. We met with Bob Zemeckis, the director of the Warner Brothers movie *Contact*, based on my novel, for which Annie and I had written a script and were now coproducing. We began negotiating on some new television and movie projects. I participated in the early stages of the encounter with Jupiter of the *Galileo* spacecraft.

But if there was one lesson I keenly learned, it is that the future is unpredictable. As William John Rogers, cheerfully penciling his postcard in the brisk air of the North Atlantic, ruefully discovered, there is no telling what even the immediate future holds. And so, after being home for months — my hair growing back, my weight back to normal, my white and red cell counts in the normal range, and me feeling absolutely splendid — another routine blood test took the wind out of my sails.

"I'm afraid I have some bad news for you," the physician said. My bone marrow had revealed the presence of a new population of dangerous, rapidly reproducing cells. In two days, the whole family was back in Seattle. I'm writing this chapter from my hospital bed at the Hutch. Through a new experimental procedure, it was determined that these anomalous cells lack an enzyme that would protect them from two standard chemotherapeutic agents — chemicals I hadn't been given before. After one round with these agents, no anomalous cells could be found in my marrow. To mop up any stragglers (they can be few but very fast-growing), I've had two more rounds of chemotherapy — to be topped off with some more cells from my sister. Once more, it seemed, I had a real shot at a complete cure.

We all have a tendency to succumb to a state of despair about the destructiveness and shortsightedness of the human species. I've certainly done my share (and on grounds I still consider well-based). But one of the discoveries of my illness is the extraordinary community of goodness to which people in my situation owe their lives.

There are more than 2 million Americans in the National Marrow Donor Program's volunteer registry, all willing to submit to a somewhat uncomfortable marrow extraction to benefit some unrelated perfect stranger. Millions more contribute blood to the American Red Cross and other blood donor institutions for no financial reward, not even a five-dollar bill, to save an unknown life.

Scientists and technicians work for years — against long odds, often for low salaries, and never with a guarantee of success. They have many motivations, but one of them is the hope of helping others, of curing diseases, of staving off death. When too much cynicism threatens to engulf us, it is buoying to remember how pervasive goodness is.

Five thousand people prayed for me at an Easter service at the Cathedral of St. John the Divine in New York City, the largest church in Christendom. A Hindu priest described a large prayer vigil for me held on the banks of the Ganges. The Imam of North America told me about his prayers for my recovery. Many Christians and Jews wrote me to tell about theirs. While I do not think that, if there is a god, his plan for me will be altered by prayer, I'm more grateful than I can say to those — including so many whom I've never met — who have pulled for me during my illness.

Many of them have asked me how it is possible to face death without the certainty of an afterlife. I can only say it hasn't been a problem. With reservations about "feeble souls," I share the view of a hero of mine, Albert Einstein:

> I cannot conceive of a god who rewards and punishes his creatures or has a will of the kind that we experience in ourselves. Neither can I nor would I want to conceive of an individual that survives his physical death; let feeble souls, from fear or absurd egotism, cherish such thoughts. I am satisfied

with the mystery of the eternity of life and a glimpse of the marvelous structure of the existing world, together with the devoted striving to comprehend a portion, be it ever so tiny, of the Reason that manifests itself in nature.

POSTSCRIPT

Since writing this chapter a year ago, much has happened. I was released from the Hutch, we returned to Ithaca, but after a few months the disease recurred. It was much more grueling this time — perhaps because my body was weakened by the previous therapies, but also because this time the pretransplant conditioning involved whole body X-irradiation. Again, my family accompanied me to Seattle. Again, I received the same expert and compassionate care at the Hutch. Again, Annie was magnificent in encouraging me and keeping my spirits up. Again, my sister, Cari, was unstintingly generous with her bone marrow. Again, I was surrounded by a community of goodness. At the moment I write — although perhaps it will have to be changed in proof — the prognosis is the best it could possibly be: All detectable bone marrow cells are donor cells, XX, female cells, cells from my sister. Not one is XY, host cells, male cells, cells that nurtured the original disease. People survive years even with a few percent of their host cells. But I won't be reasonably sure until a couple of years have passed. Until then, I can only hope.

Seattle, Washington
Ithaca, New York
October 1996

EPILOGUE

With characteristic optimism in the face of harrowing ambiguity, Carl writes the final entry in a prodigious, passionate, daringly transdisciplinary, and astonishingly original body of work.

Mere weeks later, in early December, he sat at our dinner table, regarding a favorite meal with a look of puzzlement. It held no appeal. In the best of times, my family had always prided itself on what we call "wodar," an inner mechanism that ceaselessly scans the horizon for the first blips of looming disaster. During our two years in the valley of the shadow, our wodar had remained at a constant state of highest alert. On this roller coaster of hopes dashed, raised, and dashed again, even the slightest variation in a single particular of Carl's physical condition would set alarm bells blaring.

A beat of a look passed between us. I immediately began spinning a benign hypothesis to explain away this sudden lack of appetite. As usual I was arguing that this might have nothing to do with his illness.

It was merely a fleeting disinterest in food that a healthy person might not even notice. Carl managed a little smile and just said, "Maybe." But from that moment on he had to force himself to eat and his strength declined noticeably. Despite this, he insisted on fulfilling a long-standing commitment to give two public lectures later that week in the San Francisco Bay area. When he returned to our hotel after the second talk, he was exhausted. We called Seattle.

The doctors urged us to come back to the Hutch immediately. I dreaded having to tell Sasha and Sam that we would not be returning home to them the next day as promised; that instead we would be making yet a fourth trip to Seattle, a place that had become to us synonymous with dread. The kids were stunned. How could we convincingly calm their fears that this might turn out, as it had three times before, to be another six-month stint away from home or, as Sasha immediately suspected, something far worse? Once again I went into my cheerleading mantra: Daddy wants to live. He's the bravest, toughest man I know. The doctors are the best the world has to offer. . . . Yes, Hanukkah would have to be postponed. But once Daddy was better . . .

The next day in Seattle, an X ray revealed

that Carl had a pneumonia of unknown origin. Repeated tests failed to turn up any evidence for a bacterial, viral, or fungal culprit. The inflammation in his lungs was, perhaps, a delayed reaction to the lethal dose of radiation that he had received six months before as preparation for the last bone marrow transplant. Megadoses of steroids only compounded his suffering and failed to repair his lungs. The doctors began to prepare me for the worst. Now, when I ventured out into the hospital hallway, I encountered a whole different species of expression on the too familiar faces of the staff. They either winced with sympathy or averted their eyes. It was time for the kids to come west.

When Carl saw Sasha it seemed to effect a miraculous change in his condition. "Beautiful, beautiful, Sasha," he called to her. "You are not only beautiful, but you also have enormous gorgeousness." He told her that if he did manage to survive it would be in part because of the strength her presence had given him. And for the next several hours the hospital monitors seemed to document a turnaround. My hopes soared, but in the back of my mind I couldn't help notice that the doctors didn't share my enthusiasm. They recognized this momentary rallying for what it was, what they call an

"Indian summer," the body's brief respite before its final struggle.

"This is a deathwatch," Carl told me calmly. "I'm going to die." "No," I protested. "You're going to beat this, just as you have before when it looked hopeless." He turned to me with that same look I had seen countless times in the debates and skirmishes of our twenty years of writing together and being wildly in love. With a mixture of knowing good humor and skepticism, but as ever, not a trace of self-pity, he said wryly, "Well, we'll see who's right about this one."

Sam, now five years old, came to see his father for one last time. Although Carl was by now struggling for breath and finding it harder to speak, he managed to compose himself so as not to frighten his little son. "I love you, Sam," was all he could say. "I love you, too, Daddy," Sam said solemnly.

Contrary to the fantasies of the fundamentalists, there was no deathbed conversion, no last minute refuge taken in a comforting vision of a heaven or an afterlife. For Carl, what mattered most was what was true, not merely what would make us feel better. Even at this moment when anyone would be forgiven for turning away from the reality of our situation, Carl was unflinching. As we looked deeply into each other's eyes, it was with a shared conviction that our wondrous life together was ending forever.

It had begun in 1974 at a dinner party given by Nora Ephron in New York City. I remember how handsome Carl was with his shirtsleeves rolled up and his dazzling smile. We talked about baseball and capitalism and it thrilled me that I could make him laugh so helplessly. But Carl was married and I was committed to another man. We socialized as couples. The four of us grew closer and we began to work together. There were times when Carl and I were alone with each other and the atmosphere was euphoric and highly charged, but neither of us made any sign to the other of our true feelings. It was unthinkable.

In the early spring of 1977, Carl was invited by NASA to assemble a committee to select the contents of a phonograph record that would be affixed to each of the *Voyager 1* and *2* spacecraft. Upon completion of their ambitious reconnaissance of the outermost planets and their moons, the two spacecraft would be gravitationally expelled from the Solar System. Here was an opportunity to send a message to possible beings of other worlds and times. It could be far more complex than the plaque that Carl and Carl's wife, Linda Salzman, and astronomer Frank Drake had attached to *Pioneer 10.* That was a breakthrough, but it was essentially a license plate. The *Voyager* record would include greetings in sixty human languages and one whale language, an evolutionary audio essay, 116 pictures of life on Earth and ninety minutes of music from a glorious diversity of the world's cultures. The engineers projected a one-billion-year shelf life for the golden phonograph records.

How long is a billion years? In a billion years the continents of Earth would be so altered that we would not recognize the surface of our own planet. One thousand million years ago, the most complex life forms on Earth were bacteria. In the midst of the nuclear arms race, our future, even in the short term, seemed a dubious prospect. Those of us privileged to work on the making of the Voyager message did so with a sense of sacred purpose. It was conceivable that, Noah-like, we were assembling the ark of human culture, the only artifact that would survive into the unimaginably far distant future.

In the course of my daunting search for the single most worthy piece of Chinese music, I phoned Carl and left a message at his hotel in Tucson where he was giving a talk. An hour later the phone rang in my apartment in Manhattan. I picked it up and heard a voice say: "I got back to my room and found a message that said 'Annie called.' And I asked myself, why didn't you leave me that message ten years ago?"

Bluffing, joking, I responded lightheartedly. "Well, I've been meaning to talk to you about that, Carl." And then, more soberly, "Do you mean for keeps?"

"Yes, for keeps," he said tenderly. "Let's get married."

"Yes," I said and that moment we felt we knew what it must be like to discover a new law of nature. It was a "eureka," a moment in which a great truth was revealed, one that would be reaffirmed through countless independent lines of evidence over the next twenty years. But it was also the assumption of an unlimited liability. Once you were allowed into this wonderworld, how could you ever again be content outside of it? It was June 1, our love's Holy Day. Thereafter, anytime one of us was being unreasonable with the other, the invocation of June 1 would usually bring the offender to his or her senses.

Earlier I had asked Carl if those putative

extraterrestrials of a billion years from now could conceivably interpret the brain waves of a meditator. "Who knows? A billion years is a long, long time," was his reply. "On the chance that it might be possible why don't we give it a try?"

Two days after our life-changing phone call, I entered a laboratory at Bellevue Hospital in New York City and was hooked up to a computer that turned all the data from my brain and heart into sound. I had a one-hour mental itinerary of the information I wished to convey. I began by thinking about the history of Earth and the life it sustains. To the best of my abilities I tried to think something of the history of ideas and human social organization. I thought about the predicament that our civilization finds itself in and about the violence and poverty that make this planet a hell for so many of its inhabitants. Toward the end I permitted myself a personal statement of what it was like to fall in love.

Now Carl's fever raged. I kept kissing him and rubbing my face against his burning, unshaven cheek. The heat of his skin was oddly reassuring. I wanted to do it enough so that his vibrant, physical self would become an indelibly etched sensory memory.

I was torn between exhorting him to fight on and wanting him freed from the torture apparatus of life support and the demon that had tormented him for two years.

I called his sister, Cari, who had given so much of herself to prevent this outcome, and his grown sons, Dorion, Jeremy, and Nicholas, and grandson, Tonio. Our whole family had celebrated Thanksgiving together at our house in Ithaca just weeks before. By unanimous acclaim it had been the best Thanksgiving we'd ever had. We all came away from it with a kind of glow. There had been an authenticity and a closeness in this gathering that had given us a greater sense of unity. Now I placed the phone near Carl's ear so that he could hear, one by one, their good-byes.

Our friend writer/producer Lynda Obst rushed in from Los Angeles to be with us. Lynda was there that first enchanted evening at Nora's when Carl and I met. She had witnessed firsthand, more than anyone else, both our personal and professional collaborations. As original producer of the motion picture *Contact*, she had worked closely with us for the sixteen years it had taken to guide the project into production.

Lynda had observed that the sustained incandescence of our love exerted a kind of tyranny on those around us who have been less fortunate in their search for a soul mate. However, instead of resenting our relationship, Lynda cherished it as a mathematician would an existence theorem, something that demonstrates a thing is possible. She used to call me Miss Bliss. Carl and I especially treasured those times we spent with her, laughing, talking far into the night about science, philosophy, gossip, popular culture, everything. Now this woman who had soared with us, who had been with me on the giddy day I picked out my wedding gown, was there by our side as we said good-bye forever.

For days and nights Sasha and I had taken turns whispering into Carl's ear. Sasha told him how much she loved him and all the ways that she would find in her life to honor him. "Brave man, wonderful life," I said to him over and over. "Well done. With pride and joy in our love, I let you go. Without fear. June 1. June 1. For keeps . . ."

As I make the changes in proof that Carl feared might be necessary, his son Jeremy is upstairs giving Sam his nightly computer lesson. Sasha is in her room doing homework. The *Voyager* spacecraft, with their revelations of a tiny world graced by music and love, are beyond the outermost planets, making for the open sea of interstellar space. They are hurtling at a speed of forty thousand miles per hour toward the stars and a destiny about which we can only dream. I sit surrounded by cartons of mail from people all over the planet who mourn Carl's loss. Many of them credit him with their awakenings. Some of them say that Carl's example has inspired them to work for science and reason against the forces of superstition and fundamentalism. These thoughts comfort me and lift me up out of my heartache. They allow me to feel, without resorting to the supernatural, that Carl lives.

ANN DRUYAN
February 14, 1997
Ithaca, New York

Acknowledgments*

As always, this book has been immeasurably informed and improved by Annie Druyan's insightful comments, suggestions on content and stylistic felicity, and her writing. When I grow up, I hope to be like her.

Helpful comments on some or all of this book were supplied by many friends and colleagues; I'm most grateful to them all. Among them are David Black, James Hansen, Jonathan Lunine, Geoff Marcy, Richard Turco, and George Wetherill. Others who responded generously to our requests for information include Linden Blue of General Atomics, John Bryson of Southern California Edison, Jane Callen and Jerry Donahoe at the U.S. Department of Commerce, Punam Chuhan and Julie Rickman at the World Bank, Peter Nathanielsz of the Department of Physiology at the School

*Dr. Sagan died before he was able to complete these acknowledgments. The editors regret the omission of the names of any persons or institutions he would have mentioned had he been able to complete these remarks.

of Veterinary Medicine at Cornell, James Rachels of the University of Alabama at Birmingham, Boubacar Touré at the U.N. Food and Agriculture Organization, and Tom Welch at the U.S. Department of Energy. I thank Leslie LaRocco, Department of Modern Languages and Linguistics, Cornell University, for translation services in comparing *Parade* with *Ogonyok* versions of "The Common Enemy."

I appreciate the wisdom and support of Mort Janklow and Cynthia Cannell at Janklow & Nesbit Associates, and Ann Godoff, Harry Evans, Alberto Vitale, Kathy Rosenbloom, and Martha Schwartz at Random House.

I owe a special debt to William Barnett for meticulous transcription, research assistance, proofreading, and for steering the manuscript through its various stages of completion. Bill accomplished all this while I was battling a grave illness. That I felt I could rely on him with complete confidence was a mercy for which I am grateful. Andrea Barnett and Laurel Parker of my Cornell University office provided essential correspondence and research support. I also thank Karenn Gobrecht and Cindi Vita Vogel of Annie's office for their able assistance.

While all the material in this book is

newly revised or new, the nuclei of many chapters have been previously published in *Parade*; I thank Walter Anderson, Editor-in-Chief, and David Currier, Senior Editor, for this and for their unwavering support over the years. Parts of a few chapters have appeared in *American Journal of Physics*; *Forbes-FYI*; *Environment in Peril*, Anthony Wolbarst, ed. (Washington, D.C.: Smithsonian Institution Press) (from a talk I gave to the Environmental Protection Agency, Washington, D.C.); the *Los Angeles Times Syndicate*; and *Lend Me Your Ears: Great Speeches in History*, William Safire, ed. (New York: W. W. Norton, 1992.)

Patrick McDonnell has generously agreed to the inclusion of his sketches to illustrate the text. I am also grateful to Carson Productions Group for permission to use a photograph showing me with Johnny Carson; to Barbara Boettcher for graphic artwork; to James Hansen for permission to use graphs in Chapter 11; and to Lennart Nilsson for permission to have drawings made after his pioneering photographs of human fetuses *in utero*.

References

(a few citations and suggestions for further reading)

Chapter 1, Billions and Billions

Robert L. Millet and Joseph Fielding McConkie, *The Life Beyond* (Salt Lake City: Bookcraft, 1986).

Chapter 3, Monday-Night Hunters

Harvey Araton, "Nuggets' Abdul-Rauf Shouldn't Stand for It," *The New York Times*, March 14, 1996.

A good anecdotal summary of professional sports and its admirers is *Fans!* by Michael Roberts (Washington, D.C.: New Republic Book Co., 1976). A classic study of hunter-gatherer society is *The !Kung San* by Richard Borshay Lee (New York: Cambridge University Press, 1979). Most of the hunter-gatherer customs mentioned in this book apply to the !Kung and to many other nonmarginal hunter-gatherer cultures worldwide — before they were destroyed by civilization.

Chapter 4, The Gaze of God and the Dripping Faucet

Kumi Yoshida, *et al.*, "Cause of Blue Petal Colour," *Nature*, vol. 373, 1995, p. 291.

Chapter 9, Croesus and Cassandra

Managing Planet Earth: Readings from "Scientific American" Magazine (New York: W. H. Freeman, 1990).

A. J. McMichael, *Planetary Overload: Global Environmental Change and the Health of the Human Species* (New York: Cambridge University Press, 1993).

Richard Turco, *Earth Under Siege: Air Pollution and Global Change* (New York: Oxford University Press, 1995).

Chapter 10, A Piece of the Sky Is Missing

Eric Alterman, "Voodoo Science," *The Nation*, February 5, 1996, pp. 6-7.

Richard Benedick, *Ozone Diplomacy: New Directions in Safeguarding the Planet* (Cambridge, MA: Harvard University Press, 1991).

William Brune, "There's Safety in Numbers," *Nature*, vol. 379, 1996, pp. 486-87.

Arjun Makhijani and Kevin Gurney, *Mending the Ozone Hole* (Cambridge, MA: MIT Press, 1995).

Stephen A. Montzka, *et al.*, "Decline in the Tropospheric Abundance of Halogen for Halocarbons: Implications for Stratospheric Ozone Depletion," *Science*, vol. 272, 1996, pp. 1318-22.
F. Sherwood Rowland, "The Ozone Depletion Phenomenon," in *Beyond Discovery* (Washington, D.C.: National Academy of Sciences, 1996).
James M. Russell III *et al.*, "Satellite Confirmation of the Dominance of Chlorofluorocarbons in the Global Stratospheric Chlorine Budget," *Nature*, vol. 379, 1996, pp. 526-29.

Chapter 11, Ambush: The Warming of the World

Jack Anderson, "Lessons for Us to Learn from the Persian Gulf," *Ithaca Journal*, September 29, 1990, p. 10A.
Robert Balling, Jr., "Keep Cool About Global Warming," letter to *The Wall Street Journal*, October 16, 1995, p. A14.
Hugh W. Ellsaesser, Gregory A. Inskip, and Tom M. L. Wigley, "Apply Cold Science to a Hot Topic," separate letters to *The Wall Street Journal*, November 20, 1995.
Vivien Gornitz, "Sea-Level Rise: A Review of Recent Past and Near-Future Trends," *Earth Surface Processes and Land Forms*,

vol. 20, 1995, pp. 7-20.
James Hansen, "Climatic Change: Understanding Global Warming," in *One World*, ed. by Robert Lanza (Health Press: Santa Fe, NM, 1996).
Ola M. Johannessen, *et al.*, "The Arctic's Shrinking Sea Ice," *Nature*, vol. 376, 1995, pp. 126-27.
Richard A. Kerr, "Scientists See Greenhouse, Semiofficially," *Science*, vol. 269, 1995, p. 1657.
—-, "It's Official: First Glimmer of Greenhouse Warming Seen," *Science*, vol. 270, 1995, pp. 1565-67.
Michael MacCracken, "Climate Change: the Evidence Mounts Up," *Nature*, vol. 376, 1995, pp. 645-46.
Michael Oppenheimer, "The Big Greenhouse Is Getting Warmer," letter to *The Wall Street Journal*, October 27, 1995, p. A15.
Cynthia Rosenzweig and Daniel Hillel, "Potential Impacts of Climatic Change on Agriculture and Food Supply," *Consequences*, vol. 1, Summer 1995, pp. 23-32.
Stephen E. Schwartz and Meinrat O. Andreae, "Uncertainty in Climate Change Caused by Aerosols," *Science*, vol. 272, 1996, pp. 1121-22.
William Sprigg, "Climate Change: Doctors Watch the Forecasts," *Nature*, vol.

379, 1996, p. 582.
William K. Stevens, "A Skeptic Asks, Is It Getting Hotter, or Is It Just the Computer Model?" *The New York Times*, June 18, 1996, p. C1.
Julia Uppenbrink, "Arrhenius and Global Warming," Science, vol. 272, 1996, p. 1122.

Chapter 12, Escape from Ambush

Ghossen Asrar and Jeff Dozier, *EOS: Science Strategy for the Earth Observing System* (Woodbury, NY: American Institute of Physics Press, 1994).
Business and the Environment (Cutter Information Corp.), January 1996, p. 4.
"FAS Hosts Climate Change Conference for World Bank," FAS (Federation of American Scientists), Public Interest Report, March/April 1996.
Kennedy Graham, *The Planetary Interest*, Global Security Programme, University of Cambridge, UK, 1995.
Jeremy Leggett, ed., *Global Warming* (New York: Oxford University Press, 1990).
Thomas R. Mancini, James M. Chavez, and Gregory J. Kolb, "Solar Thermal Power Today and Tomorrow," *Mechanical Engineering*, vol. 116, 1994, pp. 74-79.
Michael Valenti, "Storing Solar Energy in

Salt," *Mechanical Engineering*, vol. 117, 1995, pp. 72-75.

Chapter 13, Religion and Science: An Alliance

Julie Edelson Halport, "Harnessing the Sun and Selling It Abroad: U.S. Solar Industry in Export Boom," *The New York Times*, June 5, 1995, p. D1.

Raimon Panikkar, University of California at Santa Barbara, at Global World Forum of Spiritual and Parliamentary Leaders, Oxford, U.K., April 1988.

Carl Sagan, *et al.*, "Preserving and Cherishing the Earth," *American Journal of Physics*, vol. 58, 1990, pp. 615-17.

Peter Steinfels, "Evangelical Group Defends Laws Protecting Endangered Species as a Modern 'Noah's Ark,' " *The New York Times*, January 31, 1996.

Chapter 14, The Common Enemy

Georgi Arbatov, *The System: An Insider's Life in Soviet Politics* (New York: Times Books, 1992).

Mikhail Heller and Aleksander M. Nekrich, translated by Phyllis B. Carlos, *Utopia in Power: The History of the Soviet Union from 1917 to the Present* (New York: Summit Books, 1986).

Chapter 15, Abortion: Is It Possible to be Both "Pro-Life" and "Pro-Choice"?

John Connery, S.J., *Abortion: The Development of the Roman Catholic Perspective* (Chicago: Loyola University Press, 1977).

M. A. England, *The Color Atlas of Life Before Birth: Normal Fetal Development*, 2nd ed. (Chicago: Yearbook Medical Publishers, Inc., 1990).

Jane Hurst, *The History of Abortion in the Catholic Church: The Untold Story* (Washington, D.C.: Catholics for a Free Choice, 1989).

Carl Sagan, *The Dragons of Eden* (New York: Random House, 1977).

Carl Sagan and Ann Druyan, *Shadows of Forgotten Ancestors: A Search for Who We Are* (New York: Random House, 1992).

Chapter 17, Gettysburg and Now

Lawrence J. Korb, "Military Metamorphosis," *Issues in Science and Technology*, Winter 1995/6, pp. 75-77.

Chapter 19, In the Valley of the Shadow

Albert Einstein, *The World as I See It* (New York: Covici Friede Publishers), 1934.

Index

The employees of Thorndike Press hope you have enjoyed this Large Print book. All our Large Print titles are designed for easy reading, and all our books are made to last. Other Thorndike Press Large Print books are available at your library, through selected bookstores, or directly from us.

For information about titles, please call:

(800) 257-5157

To share your comments, please write:

Publisher
Thorndike Press
P.O. Box 159
Thorndike, Maine 04986